T0135809

Data-driven modeling of molecular interactions at the trans-Golgi network of mammalian cells

Von der Fakultät Konstruktions-, Produktions- und Fahrzeugtechnik
der Universität Stuttgart zur Erlangung der Würde eines
Doktor-Ingenieurs (Dr.-Ing.) genehmigte Abhandlung

Vorgelegt von

Patrick M. Weber

aus Leonberg

Hauptberichter: Prof. Dr.-rer.nat. Nicole Radde
Mitberichter: Prof. Dr.-rer.nat. Monilola Olayioye
                   Prof. Dr.-rer.nat. Marc-Thorsten Hütt

Tag der mündlichen Prüfung: 15 Mai 2015

Institut für Systemtheorie und Regelungstechnik

der Universität Stuttgart

2015

D 93

Bibliographic information published by the Deutsche Nationalbibliothek

The Deutsche Nationalbibliothek lists this publication in the Deutsche
Nationalbibliografie; detailed bibliographic data are available
on the Internet at http://dnb.d-nb.de .

ISBN 978-3-8325-4023-4

Logos Verlag Berlin GmbH
Comeniushof, Gubener Str. 47,
10243 Berlin
Tel.: +49 (0)30 42 85 10 90
Fax: +49 (0)30 42 85 10 92
INTERNET: http://www.logos-verlag.de

Acknowledgements

I want to thank Prof. Dr.-rer.nat. Nicole Radde for being the main supervisor of my thesis and all my research projects at the Institute for Systems Theory and Automatic Control (IST) at the University of Stuttgart.

I want to thank Prof. Dr.-rer.nat. Monilola Olayioye from the Institute of Cell Biology and Immunology (IZI) at the University of Stuttgart for being co-supervisor of my thesis and the main supervisor of the biological and experimental aspects in all collaboration research projects between the IST and the IZI.

I want to thank Prof. Dr.-rer.nat. Marc-Thorsten Hütt, head of the Computational Systems Biology group at the Jacobs University in Bremen, for being the external co-supervisor of my thesis.

I want to thank Prof. Dr.-Ing. Frank Allgöwer head of the IST and the graduate school SimTech, for hosting and supporting my research and providing me the resources of his Institute.

I want to thank Beate Spinner and Claudia Vetter for an extraordinary kind and effective support from the administrative side.

I want to thank Dr. Angelika Hausser from the IZI for the supervision of the experimental side of the collaboration projects between the IZI and the IST.

I want to thank Mariana Hornjik for performing all the wet lab experiments in the collaboration projects between the IZI and the IST. Thank you for adapting your effective experimental work in such a creative way to fit the needs of the model-based study.

I want to thank all my colleagues at the IST from the systems biology and the control group. I enjoyed the time with you all in talks, discussions, teaching, side projects and with a glass of beer in the evening. Thank you for creating the nice atmosphere at the institute.

Herein, a special thanks goes to my long year office mate and co-author Andrei Kramer. I want to thank my further co-authors Catharina Thomaseth and Dr. Jan Hasenauer for the fruitful collaboration in multiple research projects. Next, i want to thank my student co-authors Thomas Hamm and Clemens Dingler for all the important work they have done.

I acknowledge the financial support from the German Research Foundation (DFG) (GZ:RA 1840/1-1), and from the German Research Foundation within the Cluster of Excellence in Simulation Technology (EXC 310/1) at the University of Stuttgart.

I want to thank my parents, my sister and my entire family for continuously supporting me throughout my whole life in all possible ways.

Contents

Abbreviations

Methods

ODE	-	Ordinary differential equation
MLE	-	Maximum likelihood estimate
MAP	-	Maximum a posteriori probability
MCMC	-	Markov chain Monte Carlo
PT-MCMC	-	Parallel tempering Markov chain Monte Carlo
PE	-	Parameter estimation
PDF	-	Probability density function
IID	-	Independently and identically distributed

Notation

x, y, z	-	Lower case letters: variables and random variates
Z, E	-	Capital letters: random variables
$\mathcal{E}, \mathcal{Y}, \mathcal{D}$	-	Calligraphic fonts: sets, tuppels, trippels etc.
i, j, k, r	-	Indices
$\mathcal{I}, \mathcal{J}, \mathcal{R}$	-	Calligraphic fonts: index sets
n	-	Count of elements, upper bound of summation and multiplication

Cell biology

ER	-	Endoplasmic reticulum
PM	-	Plasma Membrane
SV	-	Secretory vesicles
CGN	-	Cis-Golgi network
CC	-	Cis-cisternae
MC	-	Medial-cisternae
TC	-	Trans-cisternae
TGN	-	Trans-Golgi network
ERGIM	-	Endoplasmic reticulum-Golgi intermediate compartment
MCS	-	Membrane contact site
COPI	-	Coat protein I
COPII	-	Coat protein II
CCV	-	Clathrin coated vesicles
HEK293(T)	-	Human embryonic kidney cells 293 (T = expressing Simian-Virus large T antigen)
CHO	-	Chinese hamster overay
HeLa	-	Human cervical cancer cell line (Henrietta Lacks)
COS7	-	Monkey kidney tissue model cell line
NRK	-	Normal rat kidney epithelial cells
LY-A	-	Chinese-hamster-ovary mutant cell line
FT293	-	Flp-In T-REx 293
RAW 264.7	-	Raschke Walter 264.7 (mouse macrophage cell line)

Proteins and lipids

PKD1/2/3	-	Protein kinase D1/2/3
PI4KIIIβ	-	Phosphatidyl inositol-4-kinase III β
CERT	-	Ceramide transfer protein
PKCε/ν/η	-	Protein kinase C ε/ν/η
PI	-	Phosphatidylinositol
PI(4)P	-	Phosphatidylinositol-4-phosphate
PI(4,5)P2	-	Phosphatidylinositol-4,5-bisphosphate
VAP(A/B)	-	Vesicle associated membrane associated protein (A/B)
SMS1/2/3	-	Sphingomyelin synthase 1/2/3
PP2Cε	-	Protein phosphatase 2C ε
CKIγ2	-	Casein kinase I γ-2
DAG	-	Diacylglycerol
PC	-	Phosphatidyl choline
SM	-	Sphingomyelin
Cer	-	Ceramide
DH-ceramide	-	Dihydroceramide
ARF1	-	ADP ribosylation factor 1
OSBP	-	Oxysterol binding protein
ADP	-	Adenosine diphosphate
ATP	-	Adenosine triphosphate
GFP	-	Green fluorescent protein
GST	-	Glutatione S-transferase
PA	-	Phosphatidic acid
PLD	-	Phospholipase D
PLC(β)	-	Phospholipase C (β)
PAP	-	Phosphatidic acid phosphohydrolase
LPP	-	Lipid phosphate phosphatase
CDP	-	Cytidine diphosphate
SRM	-	Serine rich motif
S#	-	Serine residue #
pS# or P#	-	Phosphorylated serine #
CYS1/2	-	Cysteine rich domains 1/2
AC	-	Aceidic region
PH	-	Pleckstrin homology
FFAT	-	Two phenylalanine in an acidic tract
START	-	Steroidogenic acute regulatory protein
siRNA	-	Small interfering ribonucleic acid
PDBu	-	Phorbol 12,13-dibutyrate
kb NB 142-70	-	9-hydroxy-3,4-dihydro-2H-[1]-benzothiolo[2,3-f][1,4]thiazepin-5-one

Summary

In this thesis we explain the findings of a model-based systems biological study, centered around the molecular interactions of protein kinase D (PKD) and ceramide transfer protein (CERT) at the trans-Golgi network (TGN) in mammalian cells. The research comprises data-driven mathematical modeling based on ordinary differential equations (ODEs) and wet lab experiments with Human Embryonic Kidney 293 T (HEK293T) cell cultures, which were particularly designed for this study.

In Chapter 1 we give an introduction to the biology of the Golgi apparatus, especially of the TGN including a summary of past and current research results. We start with the major cellular processes that are connected to the TGN. These include for example non-vesicular lipid transfer, sphingolipid homeostasis and the formation of secretory vesicles. Since these cellular processes are connected to the status of local TGN lipids and proteins we zoom in on the molecular interaction level. Herein, we establish a connection between the more general cellular processes and the particular mechanisms of the underlying molecular level which comprises TGN key proteins, kinases and lipids. We especially focus on their regulatory interactions and feedback between them. On the protein-lipid interaction level feedback involves biochemical conversions and phosphorylation reactions, which serve as the basis of the data-driven modeling process in subsequent chapters. We close the biological introduction by presenting a table that summarizes all qualitative interactions and their corresponding literature foundation. Before we move on, we give an overview of the few existing model-based studies that are related to the TGN.

In Chapter 2 we explain the theoretical framework that is needed to understand the data-driven modeling techniques that are applied in systems biological studies. Therefore we give an introduction to ordinary differential equation (ODE) models, the model class of choice for this thesis. We introduce the concept of using deterministic ODE models to describe chemical reaction kinetics. To give an insight into the model calibration process, we introduce parameter estimation via maximum likelihood based methods. Furthermore we explain the use of stochastic error models to deal with noise in experimental data and introduce statistical learning methods. We move on to sophisticated parameter estimation techniques and introduce the theory of Bayesian model analysis. We use this statistical framework as our preferred method and explain calibration of models, model predictions and model comparison. We clarify the core idea of this method, which is the analysis of models by generating samples from probability distributions. We explain local and global Markov chain Monte Carlo methods that allow us to generate these samples. We demonstrate how uncertainty analysis and probabilistic model predictions are performed with these methods and show that they allow for an adequate comparison of models with different complexities.

In Chapter 3 we demonstrate the application of data-driven Bayesian model analysis to a biochemical interaction model. The model is designed to describe the sphingomyeline synthase 1 (SMS1) reaction at the TGN in the context of a compact literature data set. The small size of the SMS1 model facilitates understanding the modeling process, the calibration of the model and model-based predictions. The SMS1 model describes a major TGN located lipid turnover reaction, which is closely connected to the reaction network of the subsequent main model study. The major biological results of the Bayesian analysis are that including a positive feedback loop in the reaction scheme clearly improves the reproduction of the experimental observations. To validate the calibrated model we correctly predict the outcome of further experimental observations related to the TGN. The important positive feedback reaction involves TGN key-players of our main study in Chapter 4. For this reason results in Chapter 3 not only demonstrate the methodology but also highly motivate our main modeling study.

In Chapter 4 we present all findings of our main modeling study concerning the TGN regulatory interactions. The established models are based on the phosphorylation dependent interactions of protein kinase D (PKD), Phosphatidylinositol-4-kinase-IIIβ and ceramide transfer protein (CERT). We start by introducing our own wet lab experiments supporting the study, since most follow-up decisions concerning the modeling are consequences of the experimental setup. Regarding measurements, Western Blot (WB) techniques with samples from HEK293T cell cultures are employed to monitor phosphorylation signals and absolute protein abundances. To stimulate dynamic responses of the TGN protein network, perturbation experiments including ectopic expression, activation and inhibition of the proteins are accomplished. To address relevant biological questions, we propose two different models varying in the ceramide transport mechanisms from the endoplasmic reticulum (ER) to the TGN. We subsequently use the absolute protein levels and time series data to calibrate our models using Bayesian parameter estimation techniques. Model comparison allows us to identify the model with the superior data fit and therefor strengthen the evidence for the respective transport mechanism. Subsequently, we use the superior model to generate quantitative predictions of TGN fluxes and protein abundances. The simulations allow to examine the biology of the endogenous TGN and gain insights that are not possible to observe directly with state of the art measurement techniques.

Major biological findings are that PKD and CERT work together in a cooperative manner to perform ceramide transfer by forming a positive feedback regulation. Additionally, we identify active PKD to be a dominant regulator of CERT dependent ceramide transfer and discover that PKD activity to the contrary is only slightly affected by altering ceramide transfer rates. We validate our model by predicting the outcome of new experiments that were not used in the model calibration process. At the end of the chapter we discuss the potential and the limitations of our TGN regulation model. We reflect the usefulness of our model in explaining the basic biology of the TGN as well as its predictive power in advanced application scenarios. As an outlook, we propose potential model extensions for follow up projects by highlighting connected pathways such as the sphingolipid metabolism or the oxysterol binding protein network.

In Chapter 5 we briefly summarize the entire thesis and put the results in a bigger context.

Kurzfassung

In dieser Doktorarbeit erläutern wir im Detail die Ergebnisse einer modellbasierten, system-biologischen Studie, die die molekularen Wechselwirkungen von Protein Kinase D (PKD) und Ceramid-Transfer-Protein (CERT) am trans-Golgi Netzwerk (TGN) von Säugetierzellen beschreibt. Die Forschungsarbeiten umfassen datengetriebene mathematische Modellierung mit gewöhnlichen Differentialgleichungen (GDGLn) und Experimentaldaten. Nasslaborexperimente mit Human Embryonic Kidney 293 T Zellkulturen wurden speziell für diese Studie konzipiert.

In Kapitel 1 bieten wir eine Einführung in die Biologie des Golgi-Apparat, insbesondere in die des TGN, einschließlich einer Zusammenfassung über den aktuellen Forschungsstand. Wir erklären wesentliche zelluläre Prozesse, die mit dem TGN in Verbindung gebracht werden, wie den nicht-vesikulären Transfer von Lipiden, Sphingolipid-Homöostase und die Bildung von Vesikeln für die Sekretion. Dabei verbinden wir allgemeine zelluläre Prozesse mit der im speziellen zugrundeliegenden molekularen Ebene und erklären im Besonderen die Wechselwirkungen von Schlüssel-Proteinen, -Kinasen und -Lipiden. Auf der Ebene von Protein-Lipid Interaktionen bedeuten Wechselwirkungen biochemische Umwandlungen und Phosphorylierungsreaktionen, die als Basis des datengetriebenen Modellierungsprozesses in den Folgekapiteln dienen. Wir beenden die biologische Einleitung, in dem wir eine Tabelle vorlegen, die alle qualitativen Interaktionen mit ihren Quellenangaben zusammenfasst. Es folgt einen Überblick über existierende modellbasierte Studien, die in Zusammenhang mit dem TGN stehen.

In Kapitel 2 erklären wir den theoretischen Rahmen, der notwendig ist, um die datengetriebenen Modellierungsmethoden, die in systembiologischen Studien angewandt werden, zu verstehen. Aus diesem Grund bieten wir eine Einleitung zu gewöhnlichen Differentialgleichungsmodellen, welche die ausgewählte Modellklasse dieser Doktorarbeit darstellen. Wir führen das Konzept ein, deterministische GDGL-Modelle zu nutzen um chemische Reaktionskinetiken zu beschreiben. Um einen Einblick in den Modellkalibrierungsprozess zu bieten, führen wir Parameterschätzung nach dem Prinzip der Maximum-Likelihood-Methode ein. Desweiteren erklären wir stochastische Fehlermodelle, um mit Messrauschen in experimentellen Daten umzugehen und erläutern statistische Lernverfahren. Wir fahren fort mit fortgeschrittenen Parameterschätzverfahren und erklären die Theorie der Baysschen Modellanalyse. Wir nutzen dieses Bayssche Framework als bevorzugte Methode und erläutern in diesem Zusammenhang Modellkalibrierung, Modellvorhersagen und Modellvergleiche. Dabei verdeutlichen wir die Kernidee dieser Methode: die Analyse von Modellen durch Generierung von Stichproben aus Wahrscheinlichkeitsverteilungen. In diesem Zusammenhang erläutern wir lokale und globale Markov-Chain-Monte-Carlo-Verfahren, die es uns erlauben, diese Stichproben zu generieren. Wir demonstrieren, wie Unsicherheitsanalyse und probabilistische Modellvorhersagen mit dieser Methode durchgeführt werden und zeigen, dass sie einen Vergleich von Modellen unterschiedlicher Komplexität ermöglicht.

In Kapitel 3 erläutern wir die Anwendung der datengetriebenen Bayesschen Modellanalyse an einem biochemischen Interaktionsmodell. Das Modell ist konzipiert, um die Sphingomyelinsynthase 1 (SMS1) Reaktion am TGN im Zusammenhang mit einem Literaturdatensatz zu beschreiben. Die kleine Größe des SMS1-Modells erleichtert das Verständnis des Modellierungsprozesses, der Kalibrierung des Modells und modellbasierter Vorhersagen. Das SMS1-Modell beschreibt eine bedeutende lokale Lipid-Umsatzreaktion am TGN, die eng mit dem Reaktionsnetzwerk der darauffolgenden Hauptmodellstudie verbunden ist. Als entscheidendes biologisches Resultat der Baysschen Analyse ergibt sich dabei, dass die Reproduktion der Experimentalergebnisse sich deutlich verbessert, wenn eine positive Feedbackschleife in das Reaktionsschema eingebunden wird. Um das kalibrierte Modell zu validieren, sagen wir weitere experimentelle Beobachtungen im Kontext des TGN korrekt voraus. Die Reaktionen der positiven Feedbackschleife involvieren die TGN-Hauptakteure aus unserer Hauptstudie in Kapitel 4. Aus diesem Grund demonstriert Kapitel 3 nicht nur die Methodik, sondern motiviert auch in hohem Maße die Hauptmodellierungsstudie.

In Kapitel 4 präsentieren wir alle Ergebnisse unserer Haupmodellierungsstudie über die regulatorischen Interaktionen am TGN. Die erstellten Modelle basieren auf den phosphorylierungsabhängigen Interaktionen von Protein Kinase D (PKD), Phosphatidylinositol-4-Kinase-IIIβ und Ceramid-Transfer-Protein (CERT). Wir beginnen damit, die Nasslaborexperimente einzuführen, da die meisten, die Modellierung betreffenden Folgeentscheidungen, Konsequenzen des experimentellen Rahmens sind. Für die Messungen werden Western Blot (WB) Techniken und Stichproben von HEK293T Zellkulturen verwendet, um phosphorylierungsabhängige Signale und absolute Proteinmengen zu beobachten. Außerdem werden Störexperimente wie ektopische Überexpression und Aktivierung oder Inhibition von Proteinen durchgeführt, die dynamische Antworten des TGN-Proteinnetzwerkes stimulieren. Um relevante biologische Fragen zu adressieren, schlagen wir zwei verschiedene Modelle vor, die sich in den Ceramid Transportmechanismen vom endoplasmatisches Retikulum (ER) zum TGN unterscheiden. Anschließend verwenden wir die absoluten Proteinkonzentrationen und Zeitreihendaten, um unsere Modelle mittels Baysscher Parameterschätzung zu kalibrieren. Modellvergleiche erlauben es uns, das Modell mit der besseren Datenübereinstimmung zu identifizieren und daher die Evidenz für den entsprechenden Transportmechanismus zu be-stärken. Danach verwenden wir das überlegene Modell, um quantitative Vorhersagen über TGN-Reaktions-flüsse und Proteinmengen zu generieren. Diese Simulationen ermöglichen es, die Biologie des endogenen TGN zu untersuchen und Einblicke zu gewinnen, die mit derzeitigen Messtechniken nicht direkt zu erlangen sind.

Wesentliche biologische Resultate sind: PKD und CERT wirken kooperativ zusammen, indem sie eine positive Rückkopplung erzeugen und so den Ceramidtransfer ermöglichen. Des Weiteren identifizieren wir aktives PKD als einen starken Regulator von CERT abhängigem Ceramidtransfer und entdecken, dass PKD-Aktivität im Gegenzug nur geringfügig von veränderten Ceramid Transferraten abhängig ist. Wir validieren unser Modell, indem wir den Ausgang weiterer Experimente vorhersagen, die nicht im Kalibrierungsprozess verwendet wurden. Am Ende des Kapitels diskutieren wir das Potential und die Anwendungsgrenzen unseres TGN Regulationsmodells. Wir reflektieren die Verwendbarkeit unseres Modells, um die grundlegenden Regulationsmechanismen des TGN zu erklären sowie seine Prädiktionsleistung in fortgeschrittenen Anwendungen. Als Ausblick schlagen wir potentielle Modellerweiterungen für Folgeprojekte vor, wie die angrenzenden Reaktionsnetzwerke des Sphingolipidmetabolismus oder das 'Oxysterol-binding protein'-Netzwerk.

In Kapitel 5 fassen wir die gesammte Doktorarbeit in Kürze zusammen und stellen die Ergebnisse in einen größeren Zusammenhang.

Chapter 1

Biology of the trans-Golgi network of mammalian cells

1.1   The Golgi apparatus

The Golgi apparatus was first identified by Camillo Golgi, an Italian physician working as Professor of General Pathology at the University of Pavia in 1897 (Droescher, 1998). Due to the limited performance of 19th century microscopes and cell labeling techniques, the discovery of the Golgi apparatus was first questioned and thought to be a measurement artifact. In the 20th century the Golgi apparatus has moved 'from artifact to center stage' (Farquhar and Palade, 1981) of modern cell biology research. This is an easily comprehensible development since the Golgi apparatus was identified to be the *post office of the cell*. More precisely, the main task of the Golgi apparatus is to receive proteins and lipids from their production sites at the endoplasmic reticulum (ER), sort and modify them, and finally sent them to multiple other cellular locations, where they are needed for a variety of functions (Mironov and Pavelka, 2009). Hence, the Golgi apparatus is an important organelle that is responsible for scheduled supply of the right type and amount of bio-molecules for other organelles.

The Golgi apparatus is found in all plant and animal cells, with a great variation in its physiology which strongly depends on the respective task of the cell. It consists of a stack of multiple compartments that are defined by lipid double layer membranes (Alberts, 2004; Karp, 2005). The right part of Figure 1.1 depicts the location of the Golgi apparatus in the cell. Other than in the simplified image, mammalian cells feature up to 40 Golgi stacks. The first stacks are located close to the endoplasmic reticulum (ER), while the remaining build up a pile of multiple layers towards the plasma membrane. Starting from the ER into the direction of the plasma membrane the Golgi apparatus comprises the cis-Golgi network (CGN), the cis-cisternae (CC), the medial-cisternae (MC), the trans-cisternae (TC) and the trans-Golgi network (TGN). For high quality images of the Golgi apparatus based on microscopy and 3D imaging, we refer to Mogelsvang *et al.* (2004). Proteins, lipids and further metabolites mostly arrive and depart from the Golgi apparatus in different types of transport vesicles. Vesicles are small spherical compartments

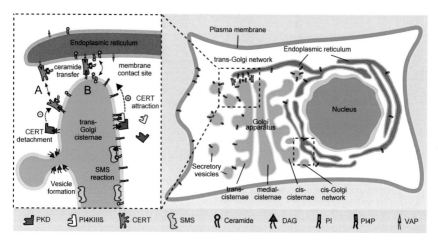

Figure 1.1: Overview of the Golgi apparatus and processes at the trans-Golgi network. *Right part:* Depiction of the orientation and location of the Golgi apparatus and its cisternae in the mammalian cell. *Inset:* Depiction of multiple processes at the TGN. The lipid transfer protein CERT establishes non-vesicular ceramide transfer between the ER and the TGN by attaching to VAP and PI4P. The proteins PKD, PI4KIIIβ interact with the membrane lipids DAG, PI and PI4P to regulate this process. The conical shaped lipid DAG forms secretory vesicles at the trans-face of the TGN cisterna by inducing membrane curvature. On the luminal facing leaflet of the TGN membrane the enzyme SMS converts ceramide into sphingomyeline.

that consist of a lipid double layer membrane that covers an aqueous solution of cargo proteins inside (Alberts, 2004). Generally, soluble cargo is stored in the interior, while lipids and other membrane affine molecules are transported in the vesicle membrane.

We now follow the way of the cargo through the Golgi apparatus: Proteins and lipids that are produced at the ER form vesicles and move along microtubuli into the ER-Golgi intermediate compartment (ERGIC), which is the site between the ER and the Golgi apparatus (Alberts, 2004; Karp, 2005). Subsequently, they fuse and form the cis-cisternae of the Golgi. These ER-to-CGN vesicles are well studied and referred to as coat protein II (COPII) vesicles (Hughes and Stephens, 2008), due to their special surface protein coating. The COPII vesicles use the dynein transport protein to move along microtubuli and are responsible for anterograde (ER to Golgi) transport.

After proteins and lipids are transported to the Golgi, they are modified via glycosylation within multiple Golgi cisternae (Alberts, 2004; Karp, 2005). This process is a very complex post translational modification and crucial for the stability or functionality of about 50% of all proteins (Varki and Chrispeels, 1999) but also essential for some lipids. Glycosylation defines the protein folding process and is an imperative necessity for the development of the correct tertiary and higher order structures of the macro molecule. The movement of the proteins through the cis-cisternae and medial-cisternae towards the trans-Golgi cisternae is complex and subject to different theories (Glick and Nakano, 2009). After traveling through the Golgi, the completely modified proteins and lipids finally arrive at the TGN. At the TGN, a not yet fully understood membrane physics process causes membrane curvature and finally forms the vesicles that destine to multiple further locations (Corda *et al.*, 2002; Shemesh *et al.*, 2003). The trans-Golgi network, particularly of mammalian cells, is the main subject of the model study within this thesis and therefore we will introduce the processes in this part of the Golgi more detailed.

1.2 The trans-Golgi network (TGN)

While the term 'trans-cisterna' usually refers to the specific part of the Golgi organelle, the term 'trans-Golgi network' is used slightly differently. The TGN covers several strongly fragmented trans-cisternae, where vesicle formation is observed. It is situated at the most distant end from the nucleus of the Golgi apparatus. These trans-cisternae membranes (or TGN membranes) are part of a protein-lipid interaction network that also involves the membranes of the ER, as we depict it in the left part of Figure 1.1. Notably, the ER itself is not a part of the TGN. Here, we illustrate the three main cellular processes at the TGN, which are important for this thesis :

1. Formation of membrane contact cites (MCS) between the trans-Golgi cisternae and the ER for non-vesicular lipid transport (NLT).

2. Lipid turnover via the sphingomyelin synthase (SMS) reaction as part of the sphingolipid pathway.

3. Transport vesicle formation at the TGN.

In the following sections we explain these important biological processes at the TGN in more detail and highlight how they are coupled with each other.

1.2.1 Formation of MCS between the trans-Golgi cisternae and the ER for non-vesicular lipid transport

Although the Golgi is generally supplied with lipids and proteins via the afore-mentioned COPII vesicles at the CGN (Funato and Riezman, 2001), lipids may also be mediated to the TGN via non-vesicular lipid transfer. As depicted in the left part of Figure 1.1, the trans-cisterna gets in close proximity with the ER, establishing so called membrane contact cites. This is possible since the ER has parts that are very distal from the nucleus, the so called rough ER. The rough ER is known to form MCSs with multiple organelles which causes the respective lipid double layers to come in close proximity with each other (< 2 nm) and exchange membrane lipids (Holthuis and Levine, 2005). A good overview of NLT processes at MCSs between multiple organelles is given in Table 1 in Lev (2010).

Two particular examples for NLT at the TGN are the transfer of ceramide from the ER to the TGN (Funato and Riezman, 2001; Hanada *et al.*, 2003) and the reverse transfer of phosphatidyl-4-phosphate from the TGN to the ER (Mesmin *et al.*, 2013). In the first case, the non-vesicular ceramide transport is mediated by a specific ceramide transport protein (CERT), as depicted in Figure 1.1 (Hanada *et al.*, 2009). In the second case, a different transfer protein named oxysterol binding protein (OSBP) is responsible for the process (Mesmin *et al.*, 2013). Non-vesicular lipid transfer is highly cargo specific, since the transfer proteins respectively need special binding sites for their lipid cargo (Lev, 2010). The reasons why this additional trafficking system is available for individual lipids are not holistically understood. However, NLT processes include a lot of regulatory interactions at the respective lipid donor and acceptor membranes, especially in the case of ceramide (Hausser *et al.*, 2005; Hanada, 2006). We now explain why the supply of the TGN via NLT of ceramide is of very high importance.

1.2.2 Sphingomyelin synthase reaction and the role of the TGN in the sphingolipid pathway

When ceramide makes its way from its synthesis sites at the ER into the TGN membrane via vesicular transfer or NLT mechanisms, it participates in one of the most important reactions of the sphingolipid pathway: the sphingomyelin synthase (SMS) reaction (Huitema *et al.*, 2004; Tafesse *et al.*, 2007). The SMS reaction is a reversible sphiongolipid/glycerolipid turnover reaction that consumes phosphatidylcholine (PC) and ceramide and produces diacylglycerol (DAG) and its eponym sphingomyelin (SM) (Huitema *et al.*, 2004). While PC is synthesized partially at the ER and in the Golgi (van Meer *et al.*, 2008), DAG and the major part of SM are synthesized in the TGN (Tafesse *et al.*, 2006) as a result of the SMS reaction. Hence, ceramide supply at the TGN is essential for the SMS reaction. Ceramide and SM belong to the class of sphingolipids and take part in an important global cellular pathway. The importance of the SMS reaction for the entire cell can be further clarified: In mammalian cells and yeast 10% to 30% of the total phospholipids for almost all cellular membranes consists of SM and above 40% consists of PC (van Meer *et al.*, 2008). The SMS reaction at the TGN can be interpreted as the entrance reaction to the sphingolipid pathway. It connects the de novo synthesis pathways of ceramide (Perry *et al.*, 2000) and PC at the ER with downstream pathways that consume e.g. SM. An example for a downstream conversion of SM, is PM and lysosome related recovery of ceramide from SM (Jenkins *et al.*, 2009). Another example for a coupling of the SMS reaction with further reactions, is the TGN related ceramide-1-phosphate production that also consumes ceramide (Hussain *et al.*, 2012).

Even though the holistic picture of the overall sphingolipid turnover is far from complete, the TGN is regarded as an important intermediate turnover station, see e.g. van Meer *et al.* (2008) and Jenkins *et al.* (2010). In summary, this clearly demonstrates the overall importance of the SMS reaction for the cell and its connection to NLT of ceramide. In the next section we introduce the connection between these two TGN processes with transport vesicle formation.

1.2.3 Transport vesicle formation at the TGN

We have introduced that COPII vesicles arrive at the Golgi from the ER and COPI vesicles maintain within Golgi transfer. We also know that e.g. SM is most exclusively produced in the TGN but consumed at different locations at the cell such as the PM (Jenkins *et al.*, 2009). This states the question how proteins and lipids leave the TGN to other destinations. The probably most popular biological process observed at the TGN is the formation of secretory transport vesicles that are destining various locations within the cell. Three types of transport vesicles can be distinguished: secretory vesicles (SV) for constitutive secretion, SVs for controlled secretion and clathrin-coated vesicles (CCV) for the selective lysosomal pathway. The three vesicle types have well defined cargo, since the glyceroprotein modifications the proteins received when passing through the Golgi subsequently serve as a sorting mechanism at the TGN. Depending on their modifications, cargo proteins accumulate at different locations of the inner side of TGN membranes. A good literature overview of modifications that can very well be connected with one of the three vesicle types is given in Traub and Kornfeld (1997). Almost all cells that have a Golgi apparatus produce the first vesicle type, constitutive SVs, as they serve to deliver membrane lipids and membrane proteins to the PM. Constitutive secretion permanently increases the amount of lipids and membrane proteins in the PM and is the primary mechanism responsible for cell growth. Controlled secretion on the other hand, is related to specialized cells that need the ability to instantly secrete a larger amount of bio-molecules into their environment. Two examples for that process are the transmitter chemicals of neurons or the digestion enzymes of pancreas cells. Other than constitutive SVs that instantly fuse with the PM, controlled SVs rest idle close to the PM in a larger group and are released upon a specific signal. The last group of vesicles are specific CCVs that deliver digestion enzymes to lysosomes. Lysosomes digest unwanted substances in the cytoplasm or internalized extracellular substances. In some cases lysosomes may subsequently be secreted themselves when they become part of the so called lysosomal secretion (Blott and Griffiths, 2002).

Aside from the specific membrane modifications that accumulate and sort the cargo, vesicle formation requires some basic membrane modifications. Among them are shifts of local lipid pools in the TGN membrane that induce membrane curvature and finally form a vesicle. Unfortunately, the scientific picture of the membrane physics behind the vesicle formation process of secretory vesicles, is incomplete. Interestingly, the lipid DAG which is a product of the SMS reaction has become of great interest in this context, since it has a conical shape and has been observed to cause negative membrane curvatures (Burger, 2000; Blott and Griffiths, 2002). Additionally, first membrane fission models that are based on DAG are able to reproduce the exact dimensions of special tubular vesicle formation processes at the TGN (Shemesh *et al.*, 2003). These results furthermore strongly connect the vesicle formation process with the SMS reaction and NLT of ceramide.

We now continue with specifying the TGN key lipids and proteins that connect these processes and introduce all relevant regulatory interactions between them.

1.3   Important TGN key-players

After having summarized the main functions of the Golgi apparatus and especially the TGN in the cellular context, we have a more detailed look at the key-players that are involved in these processes on molecular level. To optimally support our subsequent model studies, we focus on the reactants of the SMS reaction and the regulators of ceramide transfer at the ER-TGN MCS. These two processes involve a range of proteins and lipids that are related to the cytosol at ER-TGN MCSs, as well as ER and TGN membranes. The literature research presented in this chapter provides additional background information about these reactants. Furthermore we exactly explain the localization of the bio-molecules at the TGN and the biochemical reactions among the key-players or with additional molecular species. This comprises membrane binding processes, conversion reactions and covalent enzyme modifications. The experimental findings reviewed in this section are elaborated with varying cell types and experimental conditions. We always mention the cell line and give a short description of the type of experiment that supports a result to better assess the relevance of an observation for our system. This is especially important because our model in Section 4 is based on data from HEK293T cells. Since the biological picture that is established in this section strongly depends on the homology of the TGN in the organisms that were used in the experiments, we present mostly results from mammalian cell lines, that are as similar as possible to our final HEK293T cells.

At the end of this chapter we provide an overview of all experimental findings in Table 1.1. Parts of the information in this chapter is based on the literature research described in the Diploma Thesis of Thomas Michael Hamm 'Sekretionskontrolle am trans-Golgi-Netzwerk' (Institute of System Theory and Automatic Control, University of Stuttgart, 18.02.2014) which was supervised by J. Prof. Nicole Radde and the author (Hamm, 2014).

1.3.1   TGN key-lipids

In this section we introduce key-lipids that are involved in the TGN related processes introduced in Section 1.2. First we describe the lipids involved in the SMS1 reaction, namely ceramide, phosphatidyl choline (PC), diacylglycerol (DAG) and sphingomyelin (SM). Then, we discuss further lipids that are involved in CERT regulation processes like phosphatidyl inositol (PI) and phosphatidyl inositl 4 phosphate (PI4P).

1.3.1.1   Ceramide

Ceramide is a sphingolipid consisting of a sphingosine molecule and one fatty acid. It represents the most simple form of the sphingolipids since it lags an additional head group. This implies that ceramide is also contained in form of a ceramide backbone within the more complex sphingolipids. Ceramide is mostly found at the cytoplasmic leaflet of the ER membrane where the ceramide de novo pathway is located, and at the TGN. The de novo synthesis of ceramide is initiated with palmitoyl co-enzyme A and L-serine at the ER, in a four step synthesis reaction

(Perry *et al.*, 2000). A minor fraction of ER resident ceramide is directly converted into galactosylceramide. However, the major part of ceramide synthesized at the ER is transferred to the TGN by non-vesicular, CERT dependent transport. Ceramide that arrives at the CGN via vesicular transport may also traverse the Golgi apparatus via a non-vesicular intra-Golgi transport pathway to reach the TGN (D'Angelo *et al.*, 2013). At the TGN, most ceramide is consumed by the SMS1 reaction, while a minor fraction is consumed by ceramide kinase induced conversion into ceramide-1-phosphate (Hussain *et al.*, 2012). A sub-pool of ceramide, that seems to be independent of CERT transportation, is converted into glucosylceramide (GlcCer) at the Golgi (Hanada *et al.*, 2003).

Ceramide that localizes at other membranes than the ER and the TGN mostly originates from retrograde conversions or recycling pathways and not from the de novo synthesis pathway. An example of such a conversion is the retrograde hydrolysis of sphingomyelin at the plasma membrane or the lysosome which produces ceramide (Hussain *et al.*, 2012). Most sphingolipid pathway conversions are reversible, however, sphingolipids may specifically exit the reaction network in the form of ceramide. In this exit-pathway ceramide is converted into sphingosine and sphingosine-1-phosphate, which subsequently is consumed by irreversible degradations reactions (Hannun and Obeid, 2008).

Summarizing, ceramide has a central role in the sphingolipid pathway as its de novo pathway feeds the SMS1 reaction, rendering it an important pre-curser for most sphingolipids. Ceramide is additionally synthesized via retrograde conversions and irreversibly consumed by sphingolipid exit-pathways.

1.3.1.2 Phosphatidylcholine (PC)

Phosphatidylcholine is a glycerolipid, consisting of two fatty acids, a glycerol molecule, a phosphate molecule and a choline head group. The lipid PC is well studied since it is one of the most abundant membrane components of mammalian cells. In multiple important membranes, including the PM, the ER and the Golgi membrane, a fraction of 40-50% of all phospholipids is made up of PC (van Meer *et al.*, 2008). Like ceramide, it is synthesized at the ER but also partly in the TGN (van Meer *et al.*, 2008). At the TGN, PC is converted together with ceramide into DAG and SM by the reversible SMS1 reaction. Alternatively, PC can be converted into DAG by an additional reaction. Here, PC is first converted into phosphatidic acid (PA) by phospholipase D (PLD) and subsequently PA phosphohydrolases (PAPs) or lipid phosphate phosphatases (LPPs), generate DAG from PA (Lev, 2010; Bard and Malhotra, 2006). Synthesis of PC at the TGN is possible via the cytidine diphosphate-choline (CDP-choline) pathway. Here, DAG is consumed together with CDP-choline to form PC (Carrasco and Mérida, 2007). Thus, PC and DAG build up a complex network with several alternative conversions that are bypassing the SMS1 reaction. Experimental investigations suggest that especially regulation mechanisms in the CDP-choline pathway preserve critical DAG and PC levels to ensure the membrane integrity of the Golgi apparatus (Litvak *et al.*, 2005; Bard and Malhotra, 2006).

Summarizing, PC is - like ceramide - a precursor in the early sphingolipid pathway. Additionally, multiple reactions at the TGN enable bypassing the SMS related conversions and allow for a lot of external regulation of PC levels via joint pathways.

1.3.1.3 Diacylglycerol (DAG)

Diacylglycerol is a glycero lipid consisting of a glycerol molecule and two fatty acids (van Meer et al., 2008). The glycero lipid DAG is involved in a complex interconnected lipids pathway systems in the mammalian cell (Carrasco and Mérida, 2007). Its de novo production is at the ER and involves the intermediate species phosphatidic acid (PA). Furthermore, DAG is produced via several conversions in the TGN (Tafesse et al., 2006). This includes the SMS1 reaction and the CDP-choline pathway (see previous section) (Bard and Malhotra, 2006). Additionally, local pools of DAG may be produced from PI(4,5)P2 via phospholipase C (PLC) at the PM, a process connected to signal transduction. One role of DAG, especially at the TGN, is to recruit the proteins PKD and PKC to the membrane, which increases their chance of activating each other. In this process, DAG directly activates PKC upon binding. Subsequently, active PKC activates PKD which is bound to other nearby DAG lipids (Goñi and Alonso, 1999). DAG has an important function in the formation of secretory vesicles at the TGN (Bard and Malhotra, 2006). It is observed that DAG accumulations cause negative membrane curvature due to the conical shape of the lipid which initiates vesicle budding (Goñi and Alonso, 1999; Burger, 2000). Furthermore, DAG has been integrated in membrane physics models which subsequently show that DAG domains cause membrane fission (Shemesh et al., 2003; Fernández-Ulibarri et al., 2007).

Summarizing, DAG has important regulatory interactions with PKD and PKC and a major role in the vesicle budding processes. It also acts as signaling molecule at the PM. This renders DAG a central key-lipid of the TGN and multiple other membranes.

1.3.1.4 Sphingomyelin (SM)

Sphingomyelin is a sphingolipid consisting of one fatty acid, a sphingosine molecule, a phosphate group and a choline molecule. Unlike DAG and ceramide, there is no de novo synthesis pathway for SM at the ER. Hence, synthesis of SM requires the SMS1 reaction at the TGN (Tafesse et al., 2006). In mammalian cells sphingomyelin has a share of about 10% of the total phospholipids in the membranes of the TGN and and the late endosome. Its highest abundance is observed at the PM with a fraction of about 30% of the total phospholipids. This is a consequence of the constitutive secretion process that continuously adds lipid material in form of vesicles to the PM (van Meer et al., 2008). Interestingly, SM does not only serve as membrane material but is also involved in signaling between the PM and the TGN. In this context, SM forms lipid rafts with cholesterol at the PM. External hydrolyzation of SM at the PM results in perturbation of the composition of these lipid rafts. These perturbations change the phosphorylation status of CERT at the TGN by a yet undiscovered mechanism (Kumagai et al., 2007).

Additionally, SM is part of the ceramide signaling pathway at the PM. In this pathway, SM is converted by acid SMase into ceramide in response to external stimulation (Hannun and Obeid, 2002).

Summarizing, SM is an abundant PM lipid that is involved in cell stress responses and is responsible for signaling between the PM and the TGN.

1.3.1.5 Phosphatidylinositol (PI) and Phosphatidylinositol 4 phosphate (PI4P)

Phosphatidylinositol is a glycerolipid consisting of two fatty acids, a glycerol molecule and a head group including a phosphate molecule and a myo-inositol part. The phospholipid percentage of PI is about 10% in the PM, as well as the ER, Golgi and mitochondrion membranes. Other membranes contain the lipid to a lesser extent (van Meer *et al.*, 2008). De novo synthesis of PI is situated at the ER. At the TGN membrane PI is phosphorylated by PI4KIIIβ at its myo-inositole part forming phosphatidylinositol 4-phosphate (PI4P). Additionally, PI may be phosphorylated by the other PI4-kinases at other sub-cellular compartments. The PI metabolism is based on phosphorylation into multiply phosphorylated forms and retrograde dephosphorylation back into PI. It is tightly regulated in different ways depending on the respective membrane. At the TGN, especially ARF1 and neuronal calcium sensor-1 (NCS-1) serve as attractor for the TGN related kinase PI4KIIIβ (D'Angelo *et al.*, 2008).

Phosphatidylinositol 4 phosphate is a glycerolipid that consists of two fatty acids, a glycerol molecule, a phosphate molecule, a myo-inositole molecule and an additional phosphate molecule. It is most abundant at the TGN and the PM (van Meer *et al.*, 2008). The sub-pool of PI4P at the TGN serves as CERT attractor by binding the lipid transfer proteins PH domain (Hanada *et al.*, 2009). In this context, PI4P has been tested to be essential for effective ceramide transport from the ER to the TGN (Tóth *et al.*, 2006). Furthermore, PI4P attracts other lipid transfer proteins that possess a PH domain, e.g. OSBP.

Summarizing these results, PI4KIIIβ triggered conversion of PI to PI4P is the main mechanism of attraction of CERT to the TGN. PI serves as the substrate for the before mentioned reaction and is additionally known to be a very abundant glycerolipid at the TGN, the ER and the PM.

1.3.1.6 TGN lipid overview

Figure 1.2 summarizes the main properties of TGN lipids ceramide, PC, DAG and SM, which are involved in the SMS1 reaction. For further details we refer to Olayioye and Hausser (2012).

1.3.2 TGN key-proteins

The TGN proteins introduced in this section are protein kinase D (PKD) isoforms, Phosphatidyl-inositole-4-kinase-III β (PI4KIIIβ), ceramide transfer protein (CERT), sphingomyelin synthase 1 (SMS1), the protein kinase C (PKC) isoforms, vesicle-associated membrane protein-associated protein (VAP), protein phosphatase 2C epsilon (PP2Cε) and casein kinase I gamma-

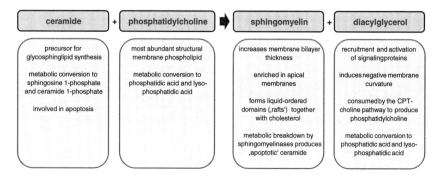

Figure 1.2: Summary of the properties of the lipids present at the TGN and involved in the SMS1 reaction. Original graphic in Olayioye and Hausser (2012).

2 (CKIγ2). For each important species, we indicate the UniProt[1] identifiers and list active domains and give information about their TGN relation. Active domains are regulatory groups within a molecule that bind or phosphorylate other TGN key-players or trigger another type of catalytic conversion.

1.3.2.1   Protein kinase D (PKD)

The members of the protein kinase D family are serine/threonine specific protein kinases (Rykx *et al.*, 2003; Rozengurt *et al.*, 2005; Wang, 2006). Three isoforms of PKD have been identified: 'protein kinase D1' (PKD1/PKCμ, UniProt identifier: Q15139, Molecular mass: 101.704 kDa, 912 amino acids), 'protein kinase D2' (PKD2, UniProt identifier: Q9BZL6, Molecular mass: 96.750 kDa, 878 amino acids) and 'protein kinase D3' (PKD3/PKCν, UniProt identifier: O94806, Molecular mass: 100,471 kDa, 890 amino acids). They all have a kinase domain responsible for their enzymatic activity and further individually differing domains (Rykx *et al.*, 2003; Wang, 2006). The regulatory domains of all three isoforms are the cystein rich domains (CYS1 and CYS2), an acidic region (AC) and a pleckstrin homology (PH) domain. There have been several studies about TGN related PKD activities of PKD1 and PKD2. Especially the role of the CYS1 domain of PKD1, the CYS2 domain of PKD2 and both of their kinase domains were discussed in the literature.

Maeda *et al.* (2001) showed that PKD1 binds with the CYS1 domain to DAG at the Golgi membrane. They observed TGN and PKD1 marker overlay in microscopy experiments and analyzed membrane and cytosolic fractions of PKD in HeLa cells. These PKD1 related results were later experimentally confirmed by Baron and Malhotra (2002). The authors successfully demonstrated *in vitro* binding between DAG and PKD1, performed an *in vivo* extraction and again observed TGN and PKD marker superposition in HeLa cells using microscopy. Furthermore,

---

[1]UniProt identifiers refer to an online archive listing human versions of the respective proteins (Consortium, 2010)

Pusapati *et al.* (2010) showed the binding of the CYS2 domain of PKD2 with ADP ribosylation factor 1 (ARF1) which is also located at the TGN membrane. In their investigations they performed extraction assays and observed superposition of respective markers via microscopy in an experiment series involving both, HeLa and HEK293T cells.

There are several literature results that show that kinase domains of PKD1 and PKD2 are able to phosphorylate PI4KIIIβ and CERT at the TGN:

Hausser *et al.* (2005) showed the colocalization of PKD and PI4KIIIβ at the TGN in HeLa and COS7 cells using microscopy experiments. Furthermore they demonstrated PKD triggered phosphorylation of PI4KIIIβ. The authors used immunoblotting to detect altered signals of phosphorylated PI4KIIIβ in HEK293T cells when silencing PKD1 and PKD2.

Fugmann *et al.* (2007) studied the colocalization of CERT and PKD1 at the TGN. The authors ectopically expressed PKD and CERT in COS7 cells and observed respective marker superposition using microscopy experiments. Furthermore they confirmed that CERT is a PKD substrate by detecting CERT using PKD-substrate antibodies in multiple experiments involving HEK293T cells. Additionally, they showed that the serine residue 132 (S132) in the CERT sequence is the phosphorylation site targeted by PKD, since mutation of serine 132 in CERT prevented detection via PKD-substrate antibodies.

Summarizing these results show that PKD1 and PKD2 bind DAG at the TGN and thereby colocalize with PI4KIIIβ and CERT in various mammalian cell lines. In this context, CERT and PI4KIIIβ are identified as PKD substrates and are phosphorylated by isoforms PKD1 and PKD2.

1.3.2.2 Phosphatidylinositol-4-kinase-IIIβ (PI4KIIIβ)

Phosphatidylinositol-4-kinase-IIIβ (PI4KIIIβ, UniProt identifier: Q9UBF8, Molecular mass: 89.807 kDa, 801 amino acids) is a cytoplasmic lipid kinase and its lipid kinase activity is strongly related to the TGN and Golgi membranes. In the previous section we already identified PI4KIIIβ as a PKD1 and PKD2 substrate. The kinase is activated due to this phosphorylation and there are additional interactions of PI4KIIIβ with other TGN related bio-molecules that have been studied in literature.

Godi *et al.* (1999) showed colocalization of PI4KIIIβ with TGN related ARF1 and ARF3 in monkey kidney tissue (COS7) cells using microscopy experiments. Additionally, they incubated Golgi membranes from normal rat kidney epithelial (NRK) cells with cytosolic solutions, that are enriched with ARF and PI4KIIIβ. The results support the hypothesis that PI4KIIIβ binds to the Golgi membrane and are consistent with their microscopy findings. Likewise, the findings of Haynes *et al.* (2007) support the hypothesis of colocalization of PI4KIIIβ and ARF1 at the TGN. They observed marker superposition in live HeLa cells using confocal microscopy.

PI4KIIIβ phosphorylates phosphtidyl inositol (PI) forming phosphatidyl-4-phosphate (PI4P) at the TGN and Golgi membranes. Godi *et al.* (1999) observed enrichment of PI4P in Golgi membranes in NRK cells after various treatments with ARF1 and PI4KIIIβ. Furthermore, a large amount of publications that investigate the role of PI4-kinases in producing PI4-phosphates are reviewed in D'Angelo *et al.* (2008).

Summarizing literature supports colocalization of PI4KIIIβ and ARF1 at the TGN in three different mammalian cell lines, while one study also supports binding with ARF3. It functions as lipid kinase which results in Golgi specific production of PI4P.

1.3.2.3   Ceramide transfer protein (CERT)

Ceramide transfer protein (CERT, UniProt identifier: Q9Y5P4, Molecular mass: 68.007kDa, 598 amino acids) is a cytoplasmic protein belonging to the class of lipid transfer proteins. The transfer protein CERT transports ceramide in a non-vesicular manner from the ER to the TGN (Hanada *et al.*, 2003). It is known that CERT has multiple functional domains that facilitate the molecular processes essential for this transfer. It binds to the ER and the TGN membranes where it extracts and releases the lipid ceramide. Additionally it allows for regulation of these actions by altering the phosphorylation of its phosphorylation sites. The respective functional groups and associated functions are:

- pleckstrin homology (PH) domain → TGN binding

- two phenylalanine in an acidic tract (FFAT) motif → ER binding

- steroidogenic acute regulatory protein (START) domain → ceramide binding

- serine repeat (SR) motif → regulation of binding affinities

The ceramide transfer process is assumed to be complex and involves the interplay of the above mentioned functional groups. The holistic picture of this process is far from complete, but there are multiple studies about the functional groups of CERT.

Some of the general properties of PH domains such as their binding with phosphoinositides, the class of lipids PI4P belongs to, is reviewed in Lemmon and Ferguson (2000). Levine and Munro (2002) investigated the TGN specific binding of CERT to PI4P by reducing the fraction of unbound CERT PH domains in PI4P titration experiments. Additionally, Hanada *et al.* (2003) studied specific binding of PH domains of CERT to PI4P at the TGN via protein-lipid overlay assays and microscopy experiments in CHO cells.

ER targeting of CERT is controlled by the FFAT motif which binds to ER resident 'vesicle-associated membrane protein-associated protein' (VAP). Kawano *et al.* (2006) demonstrated that CERT is able to bind and extract VAP, especially the isoform VAPA, from lysates of CHO cells, using immunoprecipitation experiments. A CERT mutant deficient of the FFAT motif failed to do so.

Lots of literature focuses on the START domain in CERT, especially regarding its ability to extract ceramide from membranes and binding it to CERT. A general overview of START domains binding membrane proteins is given in Alpy and Tomasetto (2005). Hanada *et al.* (2003)

showed that the START domain of CERT has the ability to extract especially ceramide from artificial phospholipid bilayers. They demonstrated that CERT is also able to transport ceramide
between donor and acceptor phospholipid vesicles. They also investigated that ceramide transport via CERT influences SM production in semi intact Chinese-hamster-ovary mutant cell line
(LY-A). These effects could not be observed for START domain deficient CERT mutants. Kumagai *et al.* (2005) showed the high specificity of CERT START domains for binding ceramide,
especially for C14-C20 ceramides, when they performed ceramide transport experiments with
artificial membranes. They also used binding assays to determine the stoichiometry of the
CERT-ceramide compound which was estimated to be 1:1. The crystallographic investigations
of Kudo *et al.* (2008) agree with these findings, and support that CERT may only bind one
ceramide molecule at a time.

The phosphorylation mechanism of the SR motif within CERT is very complex. Both the SR
phosphorylation at the TGN and the dephosphorylation at the ER have a regulatory influence on
the functionality of the PH and the START domain. The first step of the phosphorylation process
is initiated by PKD1 or PKD2 at the SR motif within CERT, particularly by phsophorylation at
the serine residue S132 (Fugmann *et al.*, 2007). Kumagai *et al.* (2007) showed that in a second
step six to eight further serine and threonine residues are phosphorylated by Casein kinase I
gamma-2 (CKIγ2), an effect known as hyperphosphorylation of the SR motif. They also report
that increasing CERT hyperphosphorylation by overexpression of CKIγ2 downregulates SM
synthesis in CHO cells.

Interestingly, the dephosphorylation process of the SR motif is connected to the ER. Saito *et al.*
(2008) published a series of successive experiments involving HEK293 cells that show, this process is triggered by ER resident protein phosphatase 2C epsilon (PP2Cε). They first observed
colocalization of PP2Cε and ER markers in microscopy experiments. In bead-based pull down
assays they then showed interactions between ER resident CERT-attractor VAPA and PP2Cε.
Finally they observed that ectopic PP2Cε expression experiments have a positive effect on
CERT dephosphorylation. Combining their results they argue that CERT is dephosphorylated
at the ER in a PP2Cε dependent manner.

Regarding the regulatory effects of the SR motif phosphorylation on PH and START domain
functioning there are several results. Kumagai *et al.* (2007) showed that hyperphosphorylation
of the SR motif causes an interaction between the START and the PH domain. They observed
that artificial phospholipid vesicles with PI4P, embedded in their membranes bind CERT to a
much lesser extend when the CERT SR motif is phosphorylated. Additionally, Tomishige *et al.*
(2009) report that hyperphosphorylation of CERT causes dissociation from the Golgi complex.
They displayed the effect in microscopy experiments in transformed human embryonal kidney
cells (FT293).

Since the experimental findings concerning the role of CERT at the TGN allow for a wide range
of conclusions we summarize them in Section 1.3.3.

1.3.2.4   ER and TGN membrane-associated proteins

To support the findings about the key-players PKD, PI4KIIIβ and CERT we shortly give an overview about the bio-molecules: 'vesicle-associated membrane protein-associated protein A', protein phosphatase 2C epsilon (PP2Cε) and Casein kinase I gamma-2 (CKIγ2). They are associated with the membrane interaction processes of PKD, PI4KIIIβ and CERT. Membrane proteins generally possess a hydrophobic terminus allowing them to integrate into membranes and a hydrophilic part, that faces towards the cytoplasm or the lumen of a compartment with functional groups to interact with additional bio-molecules.

Vesicle-associated membrane protein-associated protein A (VAPA). Vesicle-associated membrane protein-associated protein A (VAPA, UniProt identifier: Q9P0L0, Molecular mass: 27.893 kDa, 249 amino acids) belongs to the class of membrane proteins. VAPA and also VAPB have been observed to recruit cytoplasmic proteins towards the ER membrane that contain a FFAT motif (Loewen *et al.*, 2003). Besides CERT, the FFAT motif is also found in several homologs of oxysterol binding protein (OSBP), which are also known for targeting the ER and the TGN.

Protein phosphatase 2C epsilon (PP2Cε). Protein phosphatase 2C epsilon (PP2Cε, UniProt identifier: Q5SGD2, Molecular mass: 41.053 kDa, 360 amino acids) is an ER membrane resident phosphatase with a trans-membrane domain. Saito *et al.* (2008) investigated its ability to bind to VAPA via its trans-membrane domain. Together with with VAPA it serves as a CERT modifier at the ER. Thereby VAPA functions as an ER anchor for CERT and PP2Cε, which also binds VAPA and dephosphorylates the SR motif of CERT with its catalytic domain.

Casein kinase I gamma-2 (CKIγ2). Casein kinase I gamma-2 (CKIγ2, UniProt identifier: P78368, Molecular mass: 47.457 kDa, 415 amino acids) is an isoform of the casein kinases. This kinase family especially targets proteins that already have two to three phosphorylated serine or threonine residues. They are therefore responsible for several successive follow-up phosphorylations (Knippschild *et al.*, 2005). In the case of CERT, PKD acts as the initiator for the first phosphorylation of the S132 residue, while CKIγ2 continues the phosphorylation process with eight follow-up phosphorylations. The conformation changes in CERT induced by the follow-up phosphorylations finally inactivates the START domain.

1.3.2.5   Sphingomyelin synthase 1 (SMS1)

Sphingomyelin synthase 1 (SMS1, UniProt identifier: Q86VZ5, Molecular mass: 49.208 kDa, 419 amino acids) is a TGN resident membrane protein belonging to the enzyme class of phosphatidylcholine:ceramide cholinephosphotransferases. The enzyme is responsible for TGN related sphingolipid turnover as described in Section 1.2.2. SMS1 possesses several trans-membrane domains and its catalytic group is faced towards the lumen of the TGN (Huitema *et al.*, 2004). It can transfer a phospho-choline head from PC to ceramide, which results in the production of SM and DAG. The enzyme is able to catalyze the forward and backward reaction and the reaction flow can most compactly be described as

$$\text{Ceramide} + \text{PC} \underset{SMS1}{\overset{SMS1}{\rightleftharpoons}} \text{DAG} + \text{SM}. \tag{1.1}$$

There is no documented experimental evidence which direction is preferred *in vivo*. However, considering the overall picture of the TGN, it is rather possible that the flow is towards the reactants SM and DAG. The reasoning behind this assumption is that ceramide gets constantly delivered to the TGN and vesicles leaving the TGN serve as SM source for the PM and other membranes. As a consequence SM is constantly drained from the right hand side of the equation while the left hand side features a continuous ceramide inflow. A closer look at the reaction mechanism shows that six reaction steps are involved forming a reaction scheme known as substituted-enzyme mechanism (Cornish-Bowden, 2004). If we formally use the substituted-enzyme mechanism to enhance equation (1.1) it reads as

$$\text{Ceramide} + \text{PC} + \text{SMS1} \rightleftharpoons \text{Ceramide} + [\text{PC SMS1}] \rightleftharpoons \ldots \tag{1.2}$$

$$\ldots \rightleftharpoons \text{Ceramide} + [\text{DAG P}^* \text{SMS1}] \rightleftharpoons \text{Ceramide} + \text{DAG} + [\text{P}^* \text{SMS1}] \rightleftharpoons \ldots \tag{1.3}$$

$$\ldots \rightleftharpoons \text{DAG} + [\text{Ceramide P}^* \text{SMS1}] \rightleftharpoons \text{DAG} + [\text{SM SMS1}] \rightleftharpoons \ldots \tag{1.4}$$

$$\ldots \rightleftharpoons \text{DAG} + \text{SM} + \text{SMS1}. \tag{1.5}$$

First PC is integrated into the active domain of the SMS1 enzyme. Then the phospho-choline head P^* is removed from PC and binds to SMS1, creating DAG. DAG leaves the active domain of SMS1 and ceramide enters it. Ceramide receives the phospho-choline head forming SM. SM leaves the active domain of SMS1 and releases the enzyme.

In mammalian cells SMS1 is the main motor of the SM synthesis (Tafesse *et al.*, 2006, 2007), and a central part of the sphingolipid pathway, see Section 1.2.2. The reaction system is connected to the multiple other TGN reactions. Since DAG recruits PKD to the TGN it is coupled to PKD related processes. Additionally, the SMS1 reaction needs the ceramide supply generated by CERT.

Multiple studies address SMS1 related processes at the TGN. Tafesse *et al.* (2007) depleted SMS1 in HeLa cells and analyzed the effect of SMS1 activity at the TGN and the PM. In their experiments, the SMS depleted cells show a 80% reduction of SMS activity, and a reduction of cellular SM levels of about 80%. Cellular ceramide levels increased by a factor of 1.8, while PC and DAG levels stayed constant. In a further study, Villani *et al.* (2008) observed that altering SMS levels changed sub-cellular pools of DAG and changed the amount of PKD localization at the Golgi. Subathra *et al.* (2011) also depleted SMS in HeLa cells and observed changes the Golgi morphology and reduction of insulin secretion rates. Sarri *et al.* (2011) inhibited PC synthesis in CHO cells and observed a decrease in secretion rates and cellular DAG levels.

It is worth mentioning that the influence of multiple SMS isoforms is also studied in a broader cellular context beyond TGN related processes. It has multiple influences on proliferation, cell growth and survival (Tafesse *et al.*, 2007), apoptosis (Ding *et al.*, 2008), and secretion (Sarri *et al.*, 2011; Subathra *et al.*, 2011).

Summarizing these results indicate that SMS1 is involved in TGN feedback mechanisms, vesicle formation and secretion. Particularly modulation of TGN resident SMS1 or the lipids involved in the reaction influences secretory transport, TGN lipid pools and PKD-DAG interactions Subathra *et al.* (2011).

1.3.2.6   Protein kinase C η (PKCη)

Protein kinase C η (PKCη, UniProt identifier: P24723, Molecular mass: 77.828 kDa, 683 amino acids) is one of the ten existing isoforms of protein kinase C (PKC).

Several studies observe important TGN related activities od PKCη. Goodnight *et al.* (1995) and Añel and Malhotra (2005) performed immunofluorescence measurements in fibroblasts and microscopy experiments in HeLa cells and confirmed that especially the isoform PKCη localizes at the TGN. Bard and Malhotra (2006) showed that PKCη is activated at the TGN membrane, when binding to the resident lipid DAG. The authors furthermore display that PKCη overexpression significantly activates PKD1, which is also known to bind to DAG, at the TGN. Therefore, PKCη is also associated with TGN-to-PM directed vesicular transport.

In summary the experimental findings support that TGN local PKCη colocalizes with PKD at the TGN by binding DAG and subsequently activates PKD.

1.3.3   CERT related ceramide transport theories

In summary, the findings in literature support the theory that phosphorylation of CERT causes a decreased binding affinity towards PI4P at the TGN, while dephosphorylation at the ER increases the affinity towards the TGN. It is not clear if the dephosphorylation at the ER also changes CERT binding affinity to VAPA (Kumagai *et al.*, 2007; Saito *et al.*, 2008). Based on the experimental observations, the literature discusses two basic theories (A and B) describing the mechanisms of non-vesicular ceramide trafficking between the ER and the TGN (Perry and Ridgway, 2005; Hanada, 2006; Hanada *et al.*, 2007).

Theory A promotes the idea of a repeated attachment and detachment process of CERT from the ER and TGN membranes. The process is depicted in the left part of Figure 1.1. It is suggested that CERT extracts ceramide from the ER membrane, detaches from VAPA and travels short distances towards the TGN. Here, it binds to PI4P and integrates ceramide into the TGN membrane. In the next step, CERT is phosphorylated by PKD and detaches again from the TGN. Subsequently, CERT targets the ER again, binds to VAPA and repeats the sequence. The whole process forms a circular reaction scheme. There exist multiple theories about the effective distance CERT travels in this scenario (Hanada, 2010). For travel distances of 10 nm or more the authors refer to this type of CERT transport as a 'short distance shuttle'. Distances greater than 10 nm would not limit the process of CERT dependent ceramide transfer to membrane contact sites which have membrane distances of about 2nm to 10nm. However, for distances greater than 10nm the effectiveness of ER and TGN targeting is dependent on diffusion processes through the cytosol. Unfortunately there is yet no experimental data supporting this scenario. Another variant of the circular reaction suggests small travel distances ($\leq$ 10nm), comparable to spacings that are encountered at typical membrane contact sites. While this variant reduces diffusion to a minimum, a further variant of the circular reaction completely avoids travel through the cytosol at the MCS. Here, CERT remains bound to the ER and only detaches temporarily from the TGN upon phosphorylation of its SR motiv. Subsequently, it extracts new ceramide and again attaches to the TGN to deliver it.

Theory B suggests that ceramide is transferred while CERT is simultaneously binding both, the ER and the TGN membrane, as depicted in the left part of Figure 1.1. In this status CERT is transferring ceramide by repeated extraction of ceramide from the ER membrane and integration into the membrane at the TGN. This transport mechanism is referred to as 'neck-swinging' (Hanada, 2010). In this case, conformation changes and spacial restrictions within the molecular structure of the CERT protein must allow for a complex reaction sequence. The FFAT motif and the PH domain are bound to VAPA and PI4P, respectively. The START domain swings towards the ER membrane extracts ceramide and temporarily binds it. It swings towards the TGN membrane and releases the bound ceramide. Unbinding of CERT from any of the two membranes would interrupt ceramide transfer. As a consequence, CERT does not need to be phosphorylated in order to transport ceramide. More precisely, phosphorylation would prohibit the double bound status since it causes interaction of the PH and START domain, which causes detachment from the TGN membrane and finally shuts down ceramide transfer.

In summary there are some distinctive differences regarding the mechanisms between biological theories A and B. Circular reactions in theory A require CERT detachment from the TGN to transport ceramide while 'neck-swinging' in theory B does not. Hence, CERT phosphorylation has a positive effect upon ceramide transfer in all variants of theory A and an inhibitory effect in scenario B.

1.3.4   Summarizing TGN key-player interactions

Having introduced the important TGN key-players, we now summarize all literature findings that describe their regulatory interactions. Thereby we focus on interactions of the proteins PKD, PI4KIIIβ and CERT, since these proteins are part of the main models in Section 4. Results are presented in Table 1.1. The first column shows the name of the protein and the UniProt identifier. The second column explains the observed interaction or localization of the key-player. The third column lists the experiment type, cell type and the publication where the effects described in the second column have been studied.

Table 1.1: Summary of literature findings about TGN related interactions of PKD, PI4KIIIβ and CERT.

Name (UniProt)	Observation	Experiment / Cell line / Reference
PKD1 (Q15139)	specific binding to DAG at TGN	Maeda *et al.* (2001): observing superposition of PKD and TGN markers via microscopy. Comparing cytosolic and membrane fractions via Western Blot (WB). HeLa cells. Baron and Malhotra (2002): *in vitro* extraction of DAG via PKD. PKD-DAG *in vitro* binding using nitrocellulose beads. Observing superposition of PKD and TGN markers via microscopy. HeLa cells.
PKD2 (Q9BZL6)	specific binding to ARF1 at TGN	Pusapati *et al.* (2010): observing superposition of PKD2 and ARF1 markers at TGN via microscopy in HeLa cells. Bead-based extraction of ARF via PKD2 using HEK293T cells.
PKD1 & PKD2 (Q15139 & Q9BZL6)	phosphorylation of PI4KIIIβ at TGN	Hausser *et al.* (2005): observing overlay of TGN, PKD and PI4KIIIβ markers via microscopy in HeLa and COS7 cells. Detection of phospho-PI4KIIIβ after PKD silencing via WB using HEK293T cells.
PKD1 & PKD2 (Q15139 & Q9BZL6)	phosphorylation of CERT at TGN at serine S132	Fugmann *et al.* (2007): observing overlay of PKD1, CERT and TGN markers via microscopy in COS7 cells. Detection of CERT and CERT lacking S132 via PKD-substrate antibody using WBs. Monitoring phospho-CERT signals after PKD1/2 perturbations using WBs. All in HEK293 cells.
PI4KIIIβ (Q9UBF8)	localzation with ARF1 at TGN	Haynes *et al.* (2007): observing overlay of ARF, TGN and PI4KIIIβ markers with confocal microscopy in HeLa cells. Godi *et al.* (1999): observing overlay of ARF1/3, TGN and PI4KIIIβ markers via microscopy in COS7 cells.
PI4KIIIβ (Q9UBF8)	production of PI4P at TGN	Godi *et al.* (1999): detecting PI4P enrichment in Golgi membranes treated with ARF and PI4KIIIβ via thin-layer chromatography using NRK cells.
CERT (Q9Y5P4)	binding to PI4P at TGN via PH domain	Hanada *et al.* (2003) observing overlay of CERT and TGN markers via microscopy and protein-lipid overlay assays using CHO cells. Levine and Munro (2002): Monitoring unbound CERT PH domain fractions *in vitro* during PI4P titration.
CERT (Q9Y5P4)	ER targeting via FFAT motiv and binding to VAP	Kawano *et al.* (2006): extracting VAP with CERT via immunoprecipitation using CHO cells.

Table 1.1: Summary of literature findings about TGN related interactions of PKD, PI4KIIIβ and CERT.

Name (UniProt)	Observation	Experiment / Cell line / Reference
CERT (Q9Y5P4)	dephosphorylation of SR motif at ER via PP2Cε	Saito *et al.* (2008): observing overlay of PP2Cε and ER markers via microscopy in HEK293 cells. *In vitro* extraction of PP2Cε via VAPA using bead-based assays. Detection of CERT phosphorylation after PP2Cε overexpression via WB using HEK293 cells.
CERT (Q9Y5P4)	Extracting and transporting ceramide from/between membranes via START domain	Hanada *et al.* (2003): detecting ceramide extraction from phospholipid bilayers *in vitro*. Detecting transport between donor/acceptor phospholipid vesicles *in vitro*. Kumagai *et al.* (2005): identification of selective *in vitro* ceramide transport between artificial membranes. Determination of stoichiometric coefficient of CERT:ceramide via binding assays. Kudo *et al.* (2008): crystallografic analysis of CERT START domain.
CERT (Q9Y5P4)	phosphorylation of SR motif reduces affinity towards PI4P/TGN	Kumagai *et al.* (2007): detecting reduced *in vitro* binding of phosphorylated CERT to artificial vesicles containing PI4P. Tomishige *et al.* (2009): observing detachment of phosphorylated CERT from TGN via microscopy in HEK293 cells.

1.3.5 TGN key-player interactions generate feedback structures

When we summarize the interactions of the TGN key-players PKD, PI4KIIIβ and CERT from Table 1.1 we observe that the resulting biochemical reaction network forms feedback structures. At least two interrelated feedback loops can be identified. Figure 1.3 summarizes the interactions.

Ceramide is transferred from the ER to the TGN by CERT. This ceramide enhances DAG levels at the TGN via the SMS1 reaction, which in return attracts more PKC and PKD to the TGN. Colocalization of PKC and PKD subsequently activates PKD. Active PKD has two influences on CERT.

First, it directly phosphorylates and detaches CERT from the TGN, hence has a negative influence on CERT binding to the TGN (depicted by a blue arrow in Figure 1.3). Since the exact CERT transfer mechanism is unknown, this may have a positive or negative influence on the ceramide transfer rate (depicted by a green arrow in Figure 1.3), but in any case it establishes a loop closure.

Second, active PKD indirectly affects CERT via PI, PI4P and PI4KIIIβ. Active PKD activates PI4KIIIβ, which subsequently produces more PI4P from PI. Enhanced PI4P in the TGN membrane attracts CERT to the TGN, hence has a positive influence on CERT binding to the TGN . However, CERT attraction has a positive or negative influence on the ceramide transfer depending on the CERT transfer mechanism, but in any case establishes a second loop closure. The two feedback loops on the protein-lipid interaction level at the TGN render a classical analy-sis of the network very difficult. Experimental observations can yield counter-intuitive results. However, this strongly motivates model-based approaches as they are used in the field of systems or network biology (Klipp *et al.*, 2011). In the next section we reflect existing model-based studies related to the TGN.

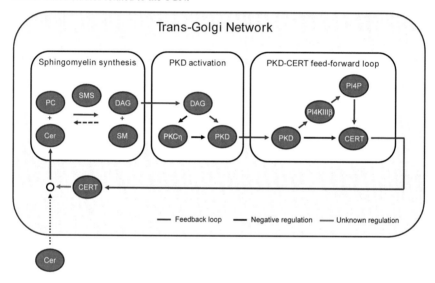

Figure 1.3: Summary of feedback mechanisms at the TGN. Red arrows follow the large feedback loop in the overall system. Adapted version of original graphic from Olayioye and Hausser (2012).

1.3.6   Existing model-based studies related to the TGN

The feedback structures at the TGN motivate a model-based approach, aiming to explain the TGN interactions with a closed loop system. To our best knowledge, there only exist four publi-cations related to the TGN that comprise mathematical modeling up to now. Gupta *et al.* (2011) uses mathematical modeling to describe a regulatory network of the sphingolipid metabolism. Hirschberg *et al.* (1998) investigate protein cargo flow through the Golgi apparatus via a math-ematical model. Two further studies propose membrane physics models (Corda *et al.*, 2002; Shemesh *et al.*, 2003). We shortly review the work of Gupta *et al.* (2011) and Hirschberg *et al.* (1998).

Gupta *et al.* (2011) introduce a sphingolipid pathway model comprising lipid conversions at multiple cellular membranes. They use ordinary differential equations (ODE) to describe sphingolipid interactions. Their model is based on lipidomics data from mouse tumor cells (RAW 264.7).

The model describes conversions of ceramide and its precursor dihydroceramide (DH-ceramide) into multiple further sphingolipids that respectively have a ceramide or a DH-ceramide backbone. Since both ceramide types are converted by the same enzymes the resulting model has a mirrored structure of two nearly identical reaction network parts. The sphingolipid and DH-sphingolipid parts of the network are connected since DH-ceramide can be converted into regular ceramide. Both model parts include the SMS1 reaction involving DAG, PC and SM. The two network parts additionally include a direct conversion from SM to ceramide bypassing the SMS1 conversion. The ceramide and the DH-ceramide subsystems both contain three identical exit pathways by irreversible conversions into ceramide phosphate, glucosyl- and galactosyl-ceramide. The DH-ceramide system has an inflow from the de novo pathway. Therefore, the reaction flow through the overall system starts at the DH-ceramide de novo pathway which feeds the DH-ceramide system which subsequently feeds the ceramide system. Boths ceramide systems feature identical reversible conversions and exit pathways. The resulting ODE model is non-linear and has 9 state variables and 25 parameters. The authors calibrate their model with lipidomics data of the before mentioned lipids from mouse tumor cells (RAW 264.7). More precisely, they study dynamic responses of the system under healthy and inflamed conditions.

The SMS1 reaction in the ceramide part of their model describes reactions located at the TGN. Therefore, their model responses and experimental observations have interesting implications with regards to the TGN. In their experiments they simultaneously increase the ceramide precursors, DAG and PC. As a result, SM levels respond immediately to the perturbations of the other SMS reaction participants (Gupta et al., 2011, Figure 2). This observation demonstrates how the SMS reaction acts as an important coupling mechanism for the lipid levels at the TGN. Hirschberg et al. (1998) established a completely different model focusing on the protein cargo flow from the ER through the Golgi towards the PM. The study focuses on monitoring GFP-tagged vesicular stomatitis virus strain ts045 G protein traffic via microscopy. The tagged protein is ectopically expressed in COS cells. Interestingly, a simple three compartment ODE model with linear kinetic rate equations is able to sufficiently explain the observed transport rates. Although the study does not focus on regulation mechanisms of secretion it gives interesting insights concerning the effectiveness of the secretory machinery. One of their major result shows that the TGN is temporarily able to send about 4000 cargo proteins per second towards the PM. They also investigated that transport rates in the presented model never showed saturation effects despite the fact that their experimental setup caused peak times when absolute abundances were high. The TGN managed to constantly send 3% of the current cargo within the Golgi to the PM within one minute. Their observations indicate that the regulation of vesicle formation is very precise and has a large potential for up-regulation of the secretion rates.

Chapter 2

Theoretical framework

In this chapter we introduce the basic theory and methods used in Chapters 3 and 4. The content comprises descriptions of the employed model class, experimental inputs, data sets and error models for experimental data as well as Bayesian parameter estimation and uncertainty quantification techniques. We conclude by presenting probabilistic model predictions and model comparison methods.

The model class chosen for the data-driven modeling in this thesis are ordinary differential equation (ODE) models. We introduce ODE models, especially their application in describing biochemical reaction networks, and explain how to simulate biological experiments via *in silico* experiment descriptions. We establish the connection between the mathematical model and the experimental data by introducing error models that add stochastic noise to deterministic model outputs. Then we establish a maximum likelihood based parameter estimation framework. We briefly address the topic of parameter identifiability to motivate that we go beyond maximum likelihood estimators to advanced Bayesian techniques. In this context we explain Markov chain Monte Carlo (MCMC) methods. We close the chapter by explaining Bayesian model predictions, uncertainty quantification, as well as model comparison using Bayes factors.

2.1 Ordinary differential equation (ODE) models for biochemical reaction networks

Ordinary differential equation models are widely used to describe biochemical interaction networks. The *BioModels Database* comprises over 600 published models, most of them being ODE models (Li *et al.*, 2010). The standard description of non-linear ODE models used in this thesis is given by

$$\mathcal{M} : \begin{cases} \dot{x}(t; u, \theta) = f(x(t; u, \theta), u(t), \theta), & x(0; u_0, \theta) = x_0(u_0, \theta), \\ y(t; u, \theta) = h(x(t; u, \theta), u(t), \theta), \end{cases} \tag{2.1}$$

with state vector $x(t; u, \theta) \in \mathbb{R}_+^{n_x}$, parameter vector $\theta \in \mathbb{R}_+^{n_\theta}$, outputs $y(t; u, \theta) \in \mathbb{R}_+^{n_y}$ and inputs $u(t) \in \mathbb{R}_+^{n_u}$, while $\mathbb{R}_+$ denotes all non-negative real numbers. State variables and parameters are always non-negative since they represent abundances, concentrations or flux rates. We assume that the vector field f is continuously differentiable to guarantee the existence of a unique solution. We allow all initial conditions to be parameter and input dependent, $x(0; u, \theta) = x_0(u, \theta)$. Since we are looking at biological regulation networks, this means that we often assume that the initial condition of the system is a flow equilibrium dependent on parameters and sometimes additional experimental inputs. Parameters in Chapters 3 and 4 are analyzed on a $\log_{10}$-scale, such that $\tilde{\theta} = 10^{diag(\theta)}$ holds. This transformation improves numerical stability when implementing the subsequently introduced parameter estimation techniques. Especially, it allows the handling of parameters with huge differences in the order of magnitude with similar numerical efficiency (Raue *et al.*, 2013).

2.2   *In silico* experiment descriptions

If an ODE model is used to simulate biological experiments, specific changes in the system inputs must be defined which represent the corresponding wet lab procedures in a correct way. Additionally, the system outputs that correspond to the measured quantities must be defined. We refer to the combination of both as an *in silico* experiment description $\mathcal{E}^i = (u^i(t), \mathcal{Y}^i)$ with $u^i(t)$ being input vectors and

$$\mathcal{Y}^i = \{y_i^j(t_k)\}_{j \in \mathcal{I}_i, k \in \mathcal{J}_i^j} \tag{2.2}$$

being a set of measurable system outputs in the experiment. In this definition the index i enumerates the different experiments and index j defines the measured output. Depending on the experiment i, the index set $\mathcal{I}_i$ may define different outputs for every individual *in silico* experiment. To ensure even more flexibility each output in the index set $\mathcal{I}_i$ has a corresponding sub-index set $\mathcal{J}_i^j$, which defines the sampling times of the respective output such that $\mathcal{J}_i^j \subseteq \{t_1, ..., t_{n_k}\}$ holds, with $t_k \in \mathbb{R}_+$. Each *in silico* experiment $\mathcal{E}^i$ corresponds to a single wet lab experiment. All $n_{\mathcal{E}}$ wet lab experiments are collected in the set $\mathcal{E} = \{\mathcal{E}^i\}_{i=1}^{n_{\mathcal{E}}}$.

2.3   Error models

In the following sections we make use of calculus with real-valued random variables and probability density functions. For a formal introduction we refer to Appendix A.

Due to the deterministic character of ODE solutions, the outcome of a system simulation cannot directly be compared to noise prone measurement data. Repeated numeric ODE integration would always return the same solution while repetitions of the experiment are influenced by measurement noise. Since comparing system outputs with measurement data is essential for inference, error models are introduced.

For the model calibration that we are going to introduce, real data is assumed to be generated by a deterministic model, e.g. an ODE describing a chemical reaction network, plus the addition of stochastic noise, the error model. Notably, the choice of the error model influences the model calibration process. Frequently used noise models in systems biology are based on independently and identically distributed (IID) random variables. Common probability distributions comprise IID additive normal noise and IID multiplicative log-normal noise (Kreutz *et al.*, 2007; Raue *et al.*, 2013). In the case additive normally distributed noise is chosen in experiment i, output j and time instant t_k, the random variable $Z_{i,j}^r(t_k)$ describing the data is defined according to

$$Z_{i,j}(t_k) = y_i^j(t_k; u^i, \theta^*) + E_{i,j,k}, \qquad (2.3)$$

with normally distributed noise $E_{i,j,k} \sim \mathcal{N}(0, \sigma^{*2}_{i,j,k})$, with mean zero and standard deviation σ^*. Here, $Z_{i,j}(t_k)$ and $E_{i,j,k}$ denote real valued random variables (see Appendix A.2). Parameter σ^* denotes the true parameter for the standard deviation and θ^* the true system parametrization. A realization of this random variable corresponding to a particular measurement unfolds as the random variate

$$z_{i,j}^r(t_k) = y_i^j(t_k; u^i, \theta^*) + e_{i,j,k,r}. \qquad (2.4)$$

The index r denotes the replicate of the measurement if the respective experiment is repeated several times.

In the case multiplicative log-normally distributed noise is chosen in experiment i, output j and time instant t_k, replicate r, the random variable $Z_{i,j}^r(t_k)$ is defined according to

$$Z_{i,j}(t_k) = y_i^j(t_k; u^i, \theta^*) \cdot E_{i,j,k}^L, \qquad (2.5)$$

with log-normally distributed noise $E_{i,j,k,r}^L \sim Log\text{-}\mathcal{N}(0, \sigma^{*2}_{i,j,k})$ with mean zero and standard deviation σ^*. A realization r unfolds as the random variate

$$z_{i,j}^r(t_k) = y_i^j(t_k; u^i, \theta^*) \cdot e_{i,j,k,r}^L. \qquad (2.6)$$

In equations (2.3) and (2.5) the random variates $z_{i,j}^r(t_k)$ represent the observed measurements and $y_i^j(t_k; u^i, \theta^*)$ denote the model outputs.

2.4 Data sets

Since experiments usually not only contain single observations $z_{i,j}^r(t_k)$, we summarize measurements in data sets. A data set comprising all measurements and replicates of a single experiment i is described by

$$\mathcal{D}^i = \{z_{i,j}^r(t_k)\}_{j \in \mathcal{I}_i, k \in \mathcal{J}_i^j, r \in \mathcal{R}_{i,j}^k}. \qquad (2.7)$$

The index set $\mathcal{R}_{i,j}^k = \{1, ..., n_{\mathrm{R}}^{j,i,k}\}$ defines the number of replicates $n_{\mathrm{R}}^{j,i,k}$ observed for experiment i, for output j of time instant t_k. The entity of the data of all performed experiments results in the complete data set defined as $\mathcal{D} = \{D^i\}_{i=1}^{n_{\mathcal{E}}}$.

We now continue to explain the model calibration, more precisely parameter estimation. For this purpose we need the biological model $\mathcal{M}$, the set of experiment descriptions $\mathcal{E}$, the data set $\mathcal{D}$ and a predefined error model.

2.5   Parameter estimation (PE)

The goal of parameter estimation is to adjust the parameters θ of a given model in a way that the model reproduces the observed data in an optimal way with regards to some predefined objective function. Therefore we need to define a quantitative measure for the quality of the model fit. A standard approach is called maximum-likelihood estimation of parameters. Maximum-likelihood estimation aims for maximizing the likelihood function, which is defined as the probability to observe the data for a particular set of model parameters. The likelihood function is constructed via the noise model. For the case of additive noise we obtain

$$E_{i,j,k} = Z_{i,j}(t_k) - y_i^j(t_k; u^i, \theta^*)\,, \tag{2.8}$$

with the normally distributed random variable $E_{i,j,k} \sim \mathcal{N}(0, \sigma^{*2}{}_{i,j,k})$. For the multiplicative log-normally distributed noise we can exploit the relation between the log-normal and the normal distribution. We use the fact that the natural logarithm of a log-normally distributed random variable $E^L \sim Log\text{-}\mathcal{N}(\mu, \sigma^2)$ is a normally distributed random variable $E \sim \mathcal{N}(\mu, \sigma^2)$ by applying a log-transformation $E = \log(E^L)$. For our error model from equation (2.5) this results in

$$E_{i,j,k}^L = \frac{Z_{i,j}(t_k)}{y_i^j(t_k; u^i, \theta^*)} \tag{2.9}$$

$$\log E_{i,j}^L = \log\left(\frac{Z_{i,j}(t_k)}{y_i^j(t_k; u^i, \theta^*)}\right) \tag{2.10}$$

$$E_{i,j,k} = \log Z_{i,j}(t_k) - \log y_i^j(t_k; u^i, \theta^*)\,. \tag{2.11}$$

After this transformation both random variables are normally distributed and the definition for the normal distribution

$$E \sim \mathcal{N}(\mu, \sigma) \tag{2.12}$$

with the probability density function (PDF)

$$f_E(e) = \frac{1}{\sigma\sqrt{2\pi}} \exp\left(\frac{-(e-\mu)^2}{2\sigma^2}\right) \tag{2.13}$$

can be used to formulate the likelihood function. In the case of IID random variables E and additive normally distributed noise, the likelihood function unfolds as a product of independent normal distributions with mean zero $\mu = 0$, and random variates $e = z_{i,j}^r(t_k) - y_i^j(t_k; u^i, \theta)$

according to

$$p(\mathcal{D}|\theta) = \prod_{i=1}^{n_{\mathcal{E}}} p(\mathcal{D}^i|\theta) \tag{2.14}$$

$$= \prod_{i=1}^{n_{\mathcal{E}}} \prod_{j \in \mathcal{I}_i} \prod_{k \in \mathcal{J}_i^j} \prod_{r \in \mathcal{R}_{i,j}^k} \frac{1}{\sqrt{2\pi}\hat{\sigma}_{i,j,k}} \exp\left\{ -\frac{1}{2} \left(\frac{z_{i,j}^r(t_k) - y_i^j(t_k; u^i, \theta)}{\hat{\sigma}_{i,j,k}} \right)^2 \right\}. \tag{2.15}$$

The estimate for the standard deviation $\hat{\sigma}_{i,j,k}$ is calculated employing the unbiased empirical variance estimate, from replicates within the data sets $\mathcal{D}^i$ from the same experiment, output and measurement time, which is defined as

$$\hat{\sigma}_{i,j,k}^2 = \frac{1}{n_R^{j,i,k} - 1} \sum_{r=1}^{n_R^{j,i,k}} (z_{i,j}^r(t_k) - \bar{z}_{i,j}^r(t_k))^2. \tag{2.16}$$

In this case $\bar{z}_{i,j}^r(t_k)$ denotes the empirical sample mean

$$\bar{z}_{i,j}^r(t_k) = \frac{1}{n_R^{j,i,k}} \sum_{r=1}^{n_R^{j,i,k}} z_{i,j}^r(t_k). \tag{2.17}$$

The likelihood function for the log-normally distributed case can be derived analogously, resulting in

$$p(\mathcal{D}|\theta) = \prod_{i=1}^{n_{\mathcal{E}}} p(\mathcal{D}^i|\theta) \tag{2.18}$$

$$= \prod_{i=1}^{n_{\mathcal{E}}} \prod_{j \in \mathcal{I}_i} \prod_{k \in \mathcal{J}_i^j} \prod_{r \in \mathcal{R}_{i,j}^k} \frac{1}{\sqrt{2\pi}\hat{\sigma}_{i,j,k}} \exp\left\{ -\frac{1}{2} \left(\frac{\log z_{i,j}^r(t_k) - \log y_i^j(t_k; u^i, \theta)}{\hat{\sigma}_{i,j,k}} \right)^2 \right\}, \tag{2.19}$$

with the empirical sample variance

$$\hat{\sigma}_{i,j,k}^2 = \frac{1}{n_R^{j,i,k} - 1} \sum_{r=1}^{n_R^{j,i,k}} (\log z_{i,j}^r(t_k) - \bar{z}_{i,j}^r(t_k))^2 \tag{2.20}$$

$$\text{and mean} \quad \bar{z}_{i,j}^r(t_k) = \frac{1}{n_R^{j,i,k}} \sum_{r=1}^{n_R^{j,i,k}} \log z_{i,j}^r(t_k). \tag{2.21}$$

To avoid numerical problems in the computation, the likelihood function is often implemented in the form of the negative log likelihood, here displayed for the example of IID additive normally distributed noise:

$$-\log p(\mathcal{D}|\theta) = \sum_{i=1}^{n_{\mathcal{E}}} p(\mathcal{D}^i|\theta) \tag{2.22}$$

$$= \sum_{i=1}^{n_{\mathcal{E}}} \sum_{j \in \mathcal{I}_i} \sum_{k \in \mathcal{J}_i^j} \sum_{r \in \mathcal{R}_{i,j}^k} -\log \frac{1}{\sqrt{2\pi}\hat{\sigma}_{i,j,k}} + \frac{1}{2} \left(\frac{z_{i,j}^r(t_k) - y_i^j(t_k; u^i, \theta)}{\hat{\sigma}_{i,j,k}} \right)^2. \tag{2.23}$$

This has the advantage that we can avoid products of small terms which cause numerical problems. If we remove all parameter-independent terms, the resulting function $J(\theta)$ coincides with the classical weighted least-squares cost functional

$$J(\theta) = = \sum_{i=1}^{n_{\mathcal{E}}} \sum_{j \in \mathcal{I}_i} \sum_{k \in \mathcal{J}_i^j} \sum_{r \in \mathcal{R}_{i,j}^k} \left(\frac{z_{i,j}^r(t_k) - y_i^j(t_k; u^i, \theta)}{\hat{\sigma}_{i,j,k}} \right)^2 . \tag{2.24}$$

Additionally, the same parameters that minimize the negative log-likelihood and the least squares functional also maximize the likelihood function according to

$$\max_{\theta} \left[p(\theta | \mathcal{D}) \right] = \min_{\theta} \left[- \log p(\theta | \mathcal{D}) \right] = \min_{\theta} \left[J(\theta) \right] . \tag{2.25}$$

Thus, the cost functional $J(\theta)$ can be used instead of $p(\theta | \mathcal{D})$ to compute maximum likelihood estimates of parameters via optimization routines. A parameter vector that maximizes the likelihood function is referred to as maximum likelihood estimate (MLE)

$$\hat{\theta}^{\mathrm{ML}} = \arg \max_{\theta} \left[p(\theta | \mathcal{D}) \right] . \tag{2.26}$$

2.6   Ill-posed problems and parameter identifiability

Minimization problems as stated in equation (2.25) are called inverse problems since the general task is to find parameters for a system or a process that reproduce a predefined behavior or some experimental observations. The converse procedure, referred to as the forward problem, would be to generate simulations or data with a known system.

Inverse problems can be classified into well-posed and ill-posed inverse problems. An inverse problem is called well-posed when it fulfills the following properties:

- existence of a solution,

- uniqueness of the solution,

- stability of the solution.

For a more technical definition we refer to Kabanikhin (2008). For inverse problems in the form of equation (2.25), a MLE $\hat{\theta}^{\mathrm{ML}}$ is generally not guaranteed to be unique, especially if the available biological data is scarce and noisy (Weber *et al.*, 2011; Gutenkunst *et al.*, 2007). If there exist multiple parameters that minimize equation (2.25), the minimization problem is ill-posed. Therefore, we introduce Bayesian methods which focus on describing the distribution of plausible parameters rather than returning only MLEs. This parameter estimation framework that can deal with ill-posed problems and allows for a probabilistic description of the solution in form of parameter samples.

2.7   Bayesian model analysis

Section 2.6 motivates a more involved approach for solving optimization problems such as eqation 2.25, since single or multiple estimates of $\hat{\theta}^{\text{ML}}$ are still insufficient to analyze the solutions of equation (2.25). Bayesian model analysis methods are statistical approaches for dealing with ill-posed inverse problems. These methods aim for a global analysis of inverse problems based on probability distributions (Gelman *et al.*, 2004). Due to the fact that Bayesian methods can deal with ill-posed inverse problems they are more and more used in the field of systems biology (Wilkinson, 2006; Eydgahi *et al.*, 2013). The main characteristic of Bayesian methods is a consistent description of all relevant quantities used in model analysis in the form of probability distributions. Bayesian methods allow to calculate parameter uncertainties, model predictions, model comparisons and model-based experiment designs using samples from probability distributions and probabilistic measures from information theory. The method employs Bayes' theorem, named after Thomas Bayes (1701-1761), who for the first time suggested the usage of the theorem in the context of updating information.

As previously mentioned, we aim on describing the distribution of plausible parameters rather than returning only MLEs. We describe our biological data z as realizations of real-valued random variables Z that are distributed according to an underlying probability density function (see Appendix A.2). If we interpret our model parameters θ as realizations of positive real valued random vectors Θ, we can apply Bayes' Theorem for probability density functions (see Appendix A.3). For a single observation z Bayes' theorem reads as

$$f_\Theta(\theta|Z = z) = \frac{f_Z(z|\Theta = \theta)f_\Theta(\theta)}{f_Z(z)} \tag{2.27}$$

or shortly

$$p(\theta|z) = \frac{p(z|\theta)p(\theta)}{p(z)}. \tag{2.28}$$

In this case $p(\theta|z)$ describes the distribution of the model parameters for the observed data z. If we furthermore want to express the distribution of the model parameters for observing the entire data set $\mathcal{D}$ we can replace $p(\theta|z)$ with the likelihood function $p(\mathcal{D}|\theta)$.

Bayes' Theorem now allows to formulate the parameter posterior distribution $p(\theta|\mathcal{D})$, describing the distribution of the model parameters, given the data $\mathcal{D}$. A straight forward application of Bayes' Theorem for random variables yields

$$p(\theta|\mathcal{D}) = \frac{p(\mathcal{D}|\theta)p(\theta)}{p(\mathcal{D})}, \tag{2.29}$$

with

- the Likelihood function $p(\mathcal{D}|\theta)$,

- the parameter prior distribution $p(\theta)$,

- the model evidence or marginal likelihood $p(\mathcal{D})$ and

- the parameter posterior distribution $p(\theta|\mathcal{D})$.

If we are aiming for calculating the posterior distribution $p(\theta|\mathcal{D})$ we need to consider the right hand side terms of equation (2.29). We already defined the likelihood function $p(\mathcal{D}|\theta)$ for our model class, see equation (2.15), hence we can focus on the remaining terms.

The prior distribution $p(\theta)$ is a data-independent probability distribution describing the prior knowledge about the parameters. The prior distribution in the Bayesian framework is sometimes compared to the upper and lower boundaries used for bounded optimization problems. However, there are generally more options for choosing a prior function, since we are allowed to describe our initial belief in terms of a parameter distribution.

We now introduce some popular priors used in the inference of biochemical reaction networks. Among them are the bounded $\log_{10}$-uniform distribution which assigns a constant probability to all parameters on an interval on a $\log_{10}$-transformed parameter scale. With the transformation $\theta = \log_{10} \tilde{\theta}$, the $\log_{10}$-uniform prior is a standard bounded uniform distribution. In the one dimensional case it reads as

$$\Theta \sim \mathcal{U}(\alpha, \beta) \tag{2.30}$$

$$\text{with } f_\Theta(\theta) = \begin{cases} \frac{1}{\beta-\alpha} & \text{for } \alpha < \theta < \beta \\ 0 & \text{elsewhere,} \end{cases} \tag{2.31}$$

with transformed upper and lower boundaries $\alpha = \log_{10} a$ and $\beta = \log_{10} b$. For Bayesian parameter inference we commonly do not need a normalized probability density function ($\int f_\Theta(\theta)d\theta = 1$) as prior function (Gelman $et\ al.$, 2004). This is especially useful for priors that are not given as an explicit function of the parameters. The non-normalized form for the prior in the multidimensional case is given by

$$f_\Theta(\theta) = \begin{cases} c & \text{for } \alpha < \theta < \beta \\ 0 & \text{elsewhere,} \end{cases} \tag{2.32}$$

where θ, α and β are vectors and $c > 0$ is a positive scalar constant.

The model evidence or marginal distribution $p(\mathcal{D})$ is a measure for the average ability of the model to reproduce the data if we integrate over all possible parameterizations. The likelihood function $p(\mathcal{D}|\theta)$ and likewise the product of the likelihood function and the prior $p(\mathcal{D}|\theta) \cdot p(\theta)$ are no normalized probability density functions with respect to the parameters. However, the integral over the domain of the parameters of the posterior distribution for a fixed data set $\mathcal{D}$ yields a value of '1', according to

$$\int_{\mathbb{R}^{n_\theta}} p(\theta|\mathcal{D})\ d\theta = 1, \tag{2.33}$$

which implies

$$p(\mathcal{D}) = \int_{\mathbb{R}^{n_\theta}} p(\mathcal{D}|\theta)p(\theta)\ d\theta. \tag{2.34}$$

Therefore, the model evidence $p(\mathcal{D})$ is also interpreted as the normalization constant of the posterior distribution. Notably, the integral of equation (2.34) can also be interpreted as the expected value of the likelihood with respect to the prior knowledge $\int p(\mathcal{D}|\theta)p(\theta)\ d\theta = E_{p(\theta)}[p(\mathcal{D}|\theta)]$, see equation (A.29) in Appendix A. Therefore, the model evidence $p(\mathcal{D})$ is used as a measure in Bayesian model comparison.

At this point it is important to mention that the evaluation of the integral of equation (2.34) is computationally expensive. It requires repeated numerical integration of the ODE system which is embedded in the likelihood function for a wide range of parameters. However, since $p(\mathcal{D})$ is a constant, the proportionality

$$p(\theta|\mathcal{D}) \propto p(\mathcal{D}|\theta)\ p(\theta) \tag{2.35}$$

holds. We exploit this proportionality in Section 2.8 to generate samples from the parameter posterior distribution $p(\theta|\mathcal{D})$ without calculating $p(\mathcal{D})$.

Parameters that maximize the posterior distribution are referred to as maximum a posteriori probability (MAP) estimates

$$\hat{\theta}_{\text{MAP}} = \arg\max_{\theta} p(\theta|\mathcal{D}) \tag{2.36}$$

in Bayesian parameter estimation.

2.8   Markov chain Monte Carlo (MCMC) sampling

As already mentioned Bayesian methods aim at describing relevant quantities in form of probability distributions. A way to describe probability distributions is to analyze representative samples from them. For this reason MCMC methods are employed to generate correlated parameter samples from $p(\theta|\mathcal{D})$. Parameter (-vector) samples can be generated by an MCMC process

$$\theta^s \overset{MCMC}{\sim} p(\theta|\mathcal{D})\,. \tag{2.37}$$

Furthermore $\mathcal{P} = \{\theta^s\}_{s=1}^{n_s}$ denotes an MCMC sample with a total number of n_s individual members θ^s. Generally, MCMC algorithms employ Markov chains to generate such a sample and successively propose a new parameter sample based on the last drawn parameter, starting from an initial point of choice (Gelman *et al.*, 2004). We now introduce the basic algorithms to generate MCMC samples.

2.8.1   Basic Metropolis Hastings sampling

One of the very first MCMC algorithms is the Metropolis Hastings algorithm (Hastings, 1970), which is still a core element of many of today's advanced sampling algorithms. The algorithm draws n_s parameter samples from the posterior distribution starting from an initial point θ^0 in the following way:

(1) Choose θ^0, such that $p(\theta^0|\mathcal{D}) > 0$, choose n_s

(2) Set s=1, set $\theta^s = \theta^0$,

(3) Draw a θ^* from a proposal distribution or transition kernel $\theta^* \sim J_k(\theta^*|\theta^s)$

(4) Calculate the ratio

$$r = \frac{p(\theta^*|\mathcal{D})}{p(\theta^s|\mathcal{D})} \cdot \frac{J_k(\theta^s|\theta^*)}{J_k(\theta^*|\theta^s)}$$

(5) Compute the acceptance rate $\alpha = \min\{r, 1\}$ and set
$$\theta^{s+1} = \begin{cases} \theta^* & \text{with probability } \alpha \\ \theta^s & \text{otherwise} \end{cases}$$

(6) END if $s \geq n_s$, otherwise set $s = s + 1$ and return to 3.

The proposal distribution $J_k(\theta^*|\theta^s)$ is preferably a probability distribution which allows for efficient generation of samples via standard random number generators. A candidate distribution is e.g. a multivariate normal distribution. In the case a symmetric proposal distribution $J_k(\theta^a|\theta^b) = J_k(\theta^b|\theta^a)$ is chosen, the algorithm is referred to as the *Metropolis* algorithm (Metropolis *et al.*, 1953). The sampling algorithms in this thesis also use exclusively symmetric proposal distributions, more precisely multivariate normal distributions, which reduces the acceptance rate calculation to

$$\alpha = \min\left\{\frac{p(\theta^*|\mathcal{D})}{p(\theta^s|\mathcal{D})}, 1\right\}. \tag{2.38}$$

The samples $\mathcal{P}$ that are generated successively by the MCMC sampling process of these algorithms have some important properties: For the Metropolis algorithm with a multivariate Gaussian proposal distribution, it is proven that the generated samples converge towards samples from the parameter posterior distribution. For an overview about the convergence details of several MCMC variants we refer to Fort *et al.* (2011). Looking at the fraction in equation (2.38) we can see, that we do not need to calculate $p(\mathcal{D})$ in order to generate MCMC samples from the posterior distribution. The constant $p(\mathcal{D})$ simply cancels out when we evaluate equation (2.38).

2.8.2 Adaptive Metropolis

The proposal distributions $J_k(\theta^*|\theta^s)$ used in the Chapters 3 and 4 are adapted throughout the sampling process, a method which is referred to as *adaptive Metropolis*. Starting from an initial multivariate normal proposal distribution $J_0(\theta^*|\theta^0) = \mathcal{N}(\theta^0, \alpha\mathcal{I})$ we generate some first samples. In this thesis we chose a small scaling $\alpha = 10^{-3}$ to avoid permanent initial rejections. At an update instant we calculate the covariance matrix of the entire sample generated up to that point. The covariance matrix serves to parametrize the new multivariate Gaussian proposal distribution. This is applied in regular intervals every few hundred samples. We also scale the proposal distribution depending on the overall acceptance rate of the proposed samples by introducing a scaling factor γ. If less then 5 % of the overall proposed samples are accepted the scaling factor γ is reduced by 10%, if more than 95% of the samples are accepted the scaling factor is increased by 10%. Initial scaling factors are set to $\gamma = 1$. The idea is to trigger a more exploratory behavior of the Markov chain in case of high acceptance rates and more conservative proposals in cases of low acceptance. This transition kernel adaptation processes can increase the convergence speed towards the true posterior distribution. It is shown that convergence towards the target distribution is still guaranteed as long as the shape adaptation processes vanishes throughout the sampling process, a property known as *diminishing adaptation* (Andrieu and Moulines, 2006). If a kernel update is performed at sample index n_u, the new transition kernel for our adaptive Metropolis scheme is defined as

$$J(\theta^*|\theta^{n_u}) = \mathcal{N}(\theta^{n_u}, \gamma \cdot Cov(\{\theta^s\}_{s=1}^{n_u})), \tag{2.39}$$

where $Cov()$ denotes the operator to calculate the covariance matrix and $\mathcal{N}(\theta^{n_u}, \gamma \cdot Cov(\{\theta^s\}_{s=1}^{n}))$ is a parametrized multivariate Gaussian distribution centered around the current sampled parameter θ^{n_u} at the update instant. This proposal distribution is used until the next update, and will always be centered around the current parameter sample θ^s of the MCMC chain, like in the original metropolis algorithm. We now introduce a strategy to improve the global mixing properties of the Markov chain.

2.8.3 Global sampling strategies: parallel tempering Markov chain Monte Carlo (PT-MCMC) sampling

To guarantee a better global mixing in the parameter sampling process, strategies have been developed to run multiple interacting Markov chains. Especially for calibrating our ODE models from Chapter 4 we slightly adapted a sampling strategy developed by Calderhead and Girolami (2009). The original idea is to run parallel interacting Markov chains that sample from the posterior distribution on different temperatures. The temperature-dependent posterior distributions, also called power posteriors, are defined as

$$p_{T_n}(\theta|\mathcal{D}) = \frac{p(\mathcal{D}|\theta)^{T_n} p(\theta)}{p_{T_n}(\mathcal{D})}, \tag{2.40}$$

$$\text{with } T_n \in [0 \ldots 1]. \tag{2.41}$$

If the temperature has a value of zero the power posterior equals the prior, while a temperature of one corresponds to the original posterior. A predefined number of chains is run in parallel, while chains are ordered according to increasing temperature values $[T_1, \ldots, T_{n_T}]$. Each chain is sampled via a Metropolis or adaptive Metropolis scheme using its individual power posterior as target distribution. An additional global strategy is implemented by allowing chains with neighboring temperatures to interact with each other by exchanging their current positions. The position exchange is a stochastic process and is embedded directly before a parameter from the proposal distribution is drawn (step (3), Section 2.8.1). In each iteration a temperature neighbor pair is drawn randomly from the discrete integer uniform distribution:

$$g \sim \mathcal{U}_\mathrm{d}(1, n_T - 1) \,, \tag{2.42}$$

with g being the index of the temperature swapping neighbor pair. The selected neighbor pair may change its current position according to the acceptance rate

$$\alpha_t = min\{r_t, 1\} \tag{2.43}$$

$$\text{with} \quad r_t = \frac{p(\mathcal{D}|\theta_{g+1}^s)^{T_g} \cdot p(\mathcal{D}|\theta_g^s)^{T_{g+1}}}{p(\mathcal{D}|\theta_g^s)^{T_g} \cdot p(\mathcal{D}|\theta_{g+1}^s)^{T_{g+1}}} \,. \tag{2.44}$$

In this context θ_g^s is the current state of the chain at temperature T_g at the current iteration s, and θ_{g+1}^s is the current state of the chain at temperature T_{g+1} respectively. All likelihood function values in equation (2.44) are already evaluated from running the parallel Metropolis schemes. Thus, temperature swapping is computationally cheap and enhances global mixing of the chains.

Our slight adaptation of Calderhead and Girolami (2009), consists in using in addition to an adaptive metropolis kernel also an acceptance rate based covariance scaling individual for each temperature, see equation (2.39).

Finally, the parameter samples from the posterior distribution serve as a basis for all further calculations needed for model analysis.

2.9   Bayesian predictions

After generating parameter samples from the posterior distribution via MCMC algorithms, these samples are used to generate model predictions in Bayesian model analysis. Model-based predictions are an important part of any model analysis framework as they offer the possibility to analyze the data fit or equivalently can be used to predict new scenarios, e.g. for model validation. Here we use a MCMC parameter sample $\{\theta^s\}_{s \in \mathcal{I}_\theta}$ from the posterior distribution as basis for uncertainty estimation. If not stated differently elsewhere, the index set $\mathcal{I}_\theta$ defines an equally spaced subsample of the entire posterior sample $\mathcal{P}$.

For model-based predictions we first define a scenario of interest i, the desired model quantities y_i with dimension n_{y_p} and the model input $u_i(t)$ corresponding to this prediction. Then we define a set of time points $t_p \in T_p = \{t_p\}_{p=1}^{n_{p_t}}$ that is of interest for the prediction. Next we generate trajectories by simulating the model with the parameter subset $\{\theta^s\}_{s \in \mathcal{I}_\theta}$ and with the new input sequence $u_i(t)$. This results in a set of model prediction vectors $\mathcal{Y}_i = \{y_i^{p,s}(t_p, \theta^s, u_i(t))\}_{s \in \mathcal{I}_\theta, t_p \in T_p}$. Depending on the output, predictions can be measurable model outputs, internal variables or new combinations of existing state variables or parameters. Inputs $u_i(t)$ can be existing or new experimental inputs. Some typical predictions to analyze biochemical regulation models are:

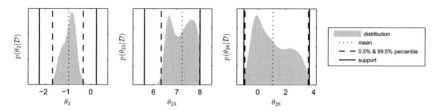

Figure 2.1: Bayesian uncertainty analysis. Shown are three examples of a one dimensional parameter uncertainty analysis. Calculations are based on a MCMC sample from Chapter 4. Depicted are the expected values, percentiles, the prior support region and a one dimensional kernel density estimate of the distribution of the respective parameter.

- Predicting outputs corresponding to measurands of training experiments and depict them together with the respective biological training data. We refer to this prediction as the Bayesian model fit.

- Predicting outcomes of biologically relevant scenarios where the respective variables cannot be measured or accessed directly. Such predictions are referred to as simulation or *in silico* studies.

- Predicting the outcome of future candidate experiments to improve experimental planning, validate or discriminate the model in successive steps.

2.10   Uncertainty analysis

In a Bayesian framework uncertainty analysis is a straight-forward procedure and can directly be applied to the parameter sample $\mathcal{P}$ or prediction samples $\mathcal{Y}_i$. To assess the uncertainty of these random variates we compute several quantities. This includes the expectation value, percentiles and an approximation of the shape of the underlying distribution denoted as kernel density estimate (KDE). For the basics of KDE computation we refer to Silverman (1986). For clarity we elaborate the uncertainty analysis along three one-dimensional examples:

A typical uncertainty analysis for parameters, calculated directly from the posterior parameter sample is depicted in Figure 2.1.

The figure depicts a parameter uncertainty analysis performed on three sub-dimensions of a 27 dimensional parameter vector, generated in the model analysis within Chapter 4. Distributions of a single or multiple, but not all sub-dimensions of a multivariate probability density function are referred to as marginal distributions, see equation (A.35) in Appendix A. Studying marginal distributions $p(\theta_i|\mathcal{D})$ is a common method in Bayesian statistics to analyze the uncertainty of single parameters.

Here, we calculated several quantities. Among them are percentiles, which give an estimate of the probability of observing values below the indicated threshold (Hyndman and Fan, 1996).

We additionally depict the boundaries of the support region of the prior distribution α and β, see equation (2.31), which in this example is a bounded log-uniform distribution. In this way we can receive information if the data restricts our parameter knowledge with respect to our prior knowledge. Based on the three examples we can distinguish several different cases: For θ_2 we gained additional information on both, the estimated upper and lower boundary from the data. Regarding θ_{23}, we only reduced one boundary. In the case of θ_{26} the distribution is restricted by the prior and the 0.5% and 99.5% percentiles are close to the boundaries of the support region. The kernel density estimate returns additional information about the underlying distribution properties like skewness (θ_2), multi modality (θ_{23}) or if the distribution is wide spread (θ_{26}). In summary, depicting the combination of expected values, percentiles and density estimates gives a good overview of the most important uncertainty properties of the underlying quantity. We apply the same uncertainty analysis for samples $\mathcal{Y}_i$ of arbitrary predictions. The respective quantities, including kernel density estimates $p(y_i(t_p)_{u_i(t)}|\mathcal{D})$, can be elaborated in an analogous way.

2.11   Model comparison

An important feature of a model analysis framework is the possibility to perform a fair comparison of models with different complexities. This especially refers to models with a different number of parameters. To gain better models it is important to compare fitting qualities and rank different hypotheses. In a Bayesian setting model comparison is based on calculating Bayes factors. Bayes factors are calculated from the model evidence term

$$p_{\mathcal{M}_A}(\mathcal{D}) = E_{p_{\mathcal{M}_A}(\theta)}[p_{\mathcal{M}_A}(\mathcal{D}|\theta)] = \int p_{\mathcal{M}_A(\theta)}(\mathcal{D}|\theta)\, p_{\mathcal{M}_A}(\theta)d\theta \; , \tag{2.45}$$

where $p_{\mathcal{M}_A}(\mathcal{D})$ stands for the model evidence, $p_{\mathcal{M}_A}(\theta)$ for the prior and $p_{\mathcal{M}_A(\theta)}(\mathcal{D}|\theta)$ for the likelihood of model $\mathcal{M}_A$ (or shortly model A). The ratio of model evidences of two different models A, B is used to calculate the Bayes factor:

$$K_{A,B} = \frac{p_{\mathcal{M}_A}(\mathcal{D})}{p_{\mathcal{M}_B}(\mathcal{D})} \; . \tag{2.46}$$

If the Bayes factor is formulated like $K_{A,B}$ it tests for superiority of model A, while $K_{B,A}$ test for superiority of model B, respectively. The superiority of a model hypothesis relative to another model can finally be determined by comparing K to published evidence tables, see Kass and Raftery (1995). If more than two models are compared, Bayes factors for all permutations of direct comparisons have to be computed and a final ranking has to be established.

Several numerical methods have been tested to approximate the computationally demanding integral in equation (2.45) (Mark and Gail, 1994; Kass and Raftery, 1995). Among these approaches especially indirect methods to estimate the integral have been proven successful, while direct methods of calculating the expected value with respect to the prior $E_{p(\theta)}[p(\mathcal{D}|\theta)]$ have been demonstrated to be rather inaccurate and error prone.

The parallel tempering MCMC method also allows for an indirect and very sophisticated approximation of $E_{p(\theta)}[p(\mathcal{D}|\theta)]$, called thermodynamic integration. At the end of a PT-MCMC run, samples from different power posteriors $p_{T_n}(\theta|\mathcal{D})$ are available. The method exploits the fact that samples of high temperature posteriors serve as excellent supporting points for the neighboring lower temperature posterior distributions. Here, we shortly summarize the method from Calderhead and Girolami (2009), for a complete derivation we refer to the original paper and references therein. Their calculations are based on an approximation of the log-marginal distribution, derived from a temperature discretization of the original integral according to

$$\log p(\mathcal{D}) \approx \frac{1}{2} \sum_{n=1}^{n_T} \Delta T_n (E_{n-1} + E_n) \,. \tag{2.47}$$

In this case ΔT_n is the temperature difference between the two neighboring temperatures of the PT-MCMC run. The expected values E_n and E_{n-1} and their approximations are defined as

$$E_n = E_{\theta|\mathcal{D},T_n}[\log p(\mathcal{D}|\theta)] \approx \frac{1}{|\mathcal{I}^{T_n}|} \sum_{i \in \mathcal{I}^{T_n}} \Delta T_n \log p(\mathcal{D}|\theta_{T_n}^i) \tag{2.48}$$

$$E_{n-1} = E_{\theta|\mathcal{D},T_{n-1}}[\log p(\mathcal{D}|\theta)] \approx \frac{1}{|\mathcal{I}^{T_{n-1}}|} \sum_{i \in \mathcal{I}^{T_{n-1}}} \Delta T_n \log p(\mathcal{D}|\theta_{T_{n-1}}^i) \,. \tag{2.49}$$

Here, $E_{\theta|\mathcal{D},T_n}$ and $E_{\theta|\mathcal{D},T_{n-1}}$ are the expectation values of the likelihood functions with respect to the power posterior distribution at temperatures T_n or T_{n-1}, and $\theta_{T_n}^i$ and $\theta_{T_{n-1}}^i$ denote samples generated from the power posteriors $p(\theta|\mathcal{D})_{T_n}$ and $p(\theta|\mathcal{D})_{T_{n-1}}$. For the Monte Carlo integration, representative subsets $\{\theta_i^{T_n}\}_{i \in \mathcal{I}_n}$ of the respective power posterior samples at temperature T_n (or respectively T_{n-1}) are employed as supporting points.

Assuming we have efficiently calculated our log-marginal distributions of two models A and B, which we want to compare, we continue with calculating the Bayes factor. The calculation of the double log-Bayes factor from the log-marginal distributions $\log p(\mathcal{D})$ unfolds as

$$2 \log K_{A,B} = 2(\log p_{\mathcal{M}_A}(\mathcal{D}) - \log p_{\mathcal{M}_B}(\mathcal{D})) \,. \tag{2.50}$$

The log-Bayes factors can also be directly compared to existing evidence tables, see Table 2.1.

Table 2.1: Evidence table for logarithmic Bayes factors (Kass and Raftery, 1995).

$2 \log K_{A,B}$	Strength of evidence
< 0	Supports model B
0 to 2	Not worth more than a bare mention
2 to 6	Positive
6 to 10	Strong
> 10	Very strong

Chapter 3

Modeling sphingomyelin synthase 1 driven conversions at the TGN

We have introduced the biological basics of the TGN key-players and their interactions, as well as the theory of Bayesian model analysis in the first two chapters. The intention of Chapter 3 is to motivate our main study in Chapter 4 and demonstrate a first application of the theoretical framework in a setting with real data. The findings presented in this section are published in Weber *et al.* (2013).

The modeling study is motivated by the observation of contra-intuitive changes in cellular lipid levels found in Ding *et al.* (2008). The authors performed ceramide, PC, SM and DAG measurements after SMS1 perturbation experiments in CHO cells and macrophages. Remember that the enzyme SMS1 is situated at the luminal side of the TGN membrane and that the SMS1 reaction is embedded in the overall regulatory network at the TGN. This renders the observations of Ding *et al.* (2008) of high relevance for understanding TGN regulation mechanisms and motivates a model-based study.

We start by introducing two simple models of the SMS1 reaction, based on chemical reaction kinetics. The first model describes the SMS1 reaction as a simple isolated enzymatic reaction, while the second model includes a positive feedback mechanism from the products to the reactants. We know that such feedback regulations do exist *in vivo* at the TGN, e.g. via protein kinase D and the ceramide transfer protein CERT. We first use Bayesian methods to calibrate both models to the data. Subsequently, we investigate if the proposed feedback extension can help to explain the experimental data. Furthermore, we connect our results to additional findings from the literature and discuss possible model extensions. Finally, we motivate our main TGN model study by suggesting a further investigation of these relevant SMS1 related feedback structures.

3.1 Experimental data

In data-driven modeling, most decisions concerning modeling are based on assessing the experimental data. Ding *et al.* (2008) measured changes of ceramide, PC, SM and DAG levels while manipulating different SMS isoforms via overexpression and knockdown experiments. We especially look at the SMS1 related experiments. A first set of experiments summarizes observations after SMS1 overexpression in CHO cells (original Table 1 in Ding *et al.* (2008)). Further experiments summarize observations after SMS1 silencing in macrophages (original Table 2 in Ding *et al.* (2008)). We normalized both data sets to the organism specific endogenous lipid levels and combined them, creating a data set that is comparable across cell lines. This aggregation and normalization of the data implies that we have to formulate our model for two organisms, assuming a qualitative homology in the organization of their SMS1 reaction. The resulting normalized data are shown in Tables 3.1 and 3.2 (Weber *et al.* (2013)).

Table 3.1: Lipid concentrations and respective standard deviations measured in CHO cells overexpressing SMS1, which resulted in a 2.2-fold increase in cellular SMS1 activity. Values according to Table 1 in Ding *et al.* (2008) and normalized to the levels in the control experiments. Values labeled with an asterix were marked as statistically significant changes compared to the control in the original publication.

Type of cell	[Cer]	[PC]	[SM]	[DAG]
Ctrl.	1	1	1	1
S.D. Ctrl $\sigma^{u=1}$	0.09	-	0.12	0.21
SMS1 overexpr.	1.44*	0.96	1.24*	1.35*
S.D. SMS1 overexpr. $\sigma^{u=2.2}$	0.08	-	0.12	0.27

Table 3.2: Lipid concentrations and respective standard deviations in SMS1 knockdown THP-1-derived macrophages, which resulted in a 23% decrease in SMS1 activity. Values according to Table 2 in Ding *et al.* (2008) are normalized to the levels in the control experiments. Values labeled with an asterix were marked as statistically significant changes compared to the control in the original publication.

Type of cell	[Cer]	[PC]	[SM]	[DAG]
Ctrl.	1	1	1	1
S.D. Ctrl $\sigma^{u=1}$	0.08	-	0.11	0.05
SMS1 knockdown	0.95	1.06	0.8*	0.76*
S.D. SMS1 knockdown $\sigma^{u=0.77}$	0.12	-	0.09	0.10

The qualitative summary of the outcome of the experiments is depicted in Table 3.3. Keeping the isolated SMS1 reaction in mind (see chemical equation (1.1) in Section (1.3.2.5)), we expected increased abundances of the products DAG and SM in the SMS1 overexpression experiments and decreased DAG and SM abundances in the silencing experiments. These expectations were confirmed by the observations.

We also expected constant PC levels due to results of multiple other publications (Tafesse *et al.*, 2007; Ding *et al.*, 2008), and again experimental observations satisfied our expectations. Furthermore, we expected that the abundance of the educt ceramide decreases with overexpression of SMS1 and increases with SMS1 silencing. However, ceramide levels fell short of expectation. In fact, the data shows increased ceramide levels for the SMS1 overexpression and unaltered values for SMS1 silencing. We used these experimental observations as a motivation to formulate two model hypothesis. One of them includes a simple linear positive feedback from DAG to ceramide. This is for sure a huge oversimplification for the *in vivo* feedback structures. However, with regards to the limited data, we first tested a minimal solution. We now elaborate on the modeling details.

Table 3.3: Comparison between experimental observations and expected responses of model 1. Shown are the qualitative changes in dynamic lipid steady state concentrations after SMS1 manipulations (Weber *et al.*, 2013).

		[Cer]		[PC]		[DAG]		[SM]	
Data		SMS1↑	SMS1↓	SMS1↑	SMS1↓	SMS1↑	SMS1↓	SMS1↑	SMS1↓
		↑	−	−	−	↑	↓	↑	↓
Expected		SMS1↑	SMS1↓	SMS1↑	SMS1↓	SMS1↑	SMS1↓	SMS1↑	SMS1↓
		↓	↑	−	−	↑	↓	↑	↓

3.2 SMS1 reaction models

As already introduced in Section 1.3.2.5, we modeled the SMS1 reaction as a simple reversible reaction (Huitema *et al.*, 2004). Additionally, we modified the reaction so that it mimics a net flow directed towards DAG and SM by adding a ceramide inflow and degradation terms to ceramide, DAG and SM. PC levels were modeled as constants as they seem to be highly robust with respect to experimental manipulations (Tafesse *et al.*, 2007; Ding *et al.*, 2008) and have a high overall abundance in TGN membranes (van Meer *et al.*, 2008). The overall system reactions can be summarized in the following way:

Table 3.4: Chemical conversion reactions of the SMS1 system.

$\emptyset \xrightarrow{C_{in}} \text{Cer}$ Influx of ceramide at the TGN

$\text{Cer} \xrightarrow{d_1} \emptyset$ Degradation of ceramide

$\text{PC} = \text{const.}$ Constant PC levels

$\text{Cer} + \text{PC} \underset{\text{SMS1}}{\overset{\text{SMS1}}{\rightleftharpoons}} \text{SM} + \text{DAG}$ SMS1 driven reversible reaction

$\text{DAG} \xrightarrow{d_2} \emptyset$ Degradation of DAG

$\text{SM} \xrightarrow{d_3} \emptyset$ Degradation of SM

The goal of the model study was to compare two versions of this system. In Figure 3.1 A we assumed that model 1 represents the reaction isolated and restricted by smaller model boundaries, with a constant inflow of ceramide. In contrast, model 2 features the biologically motivated feedback regulation (see Section 1.3.5) from DAG to ceramide involving the kinases PKD and PI4KIIIβ. Figure 3.1 B shows a graph of the feedback regulation. DAG recruits PKD to the TGN, where it is activated by PKCη (Bard and Malhotra, 2006). Active PKD influences CERT in two ways: Direct phosphorylation and detachment from the TGN and indirect attraction via activation of PI4KIIIβ (Fugmann *et al.*, 2007), which initiates PI4P production from PI. Finally, CERT delivers new ceramide to the TGN in a yet not fully understood way.

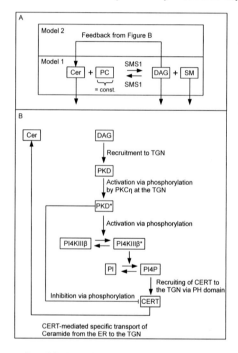

Figure 3.1: A: Structure of models 1 and 2: model 1 considers only the isolated lipids at the TGN that take part in the SMS1 reaction. Model 2 includes a positive feedback from DAG to ceramide. **B:** Illustration of the feedback regulation from DAG to ceramide. DAG recruits PKD to the TGN, where it is activated via phosphorylation by PKCη. Active PKD initiates the phosphorylation and thereby activation of the kinase PI4KIIIβ, which subsequently supports the production of PI4P from PI. CERT is attracted by PI4P to the TGN. This enables the transfer of ceramide to the TGN, where it can be converted to SM. Active PKD may also directly detach CERT from the TGN by phosphorylation.

To realize two ODE models of the SMS1 reaction we needed to decide upon the reaction kinetics that we wanted to use for the reactions in Table 3.4. In a data-driven study the model is usually kept as simple as possible: The objective is to include the given experimental data and test specific experimental scenarios without gaining too much complexity. We decided to use the following ODE system

$$
\begin{aligned}
\dot{x}_1 &= C_{in} - d_1 x_1 - v(u, \mathbf{x}, [PC]) & \text{Ceramide} \\
\dot{x}_2 &= -d_2 x_2 + v(u, \mathbf{x}, [PC]) & \text{DAG} \\
\dot{x}_3 &= -d_3 x_3 + v(u, \mathbf{x}, [PC]) & \text{SM},
\end{aligned}
\tag{3.1}
$$

with variation for the term

$$
C_{in} = \begin{cases}
s_1 & \text{constant inflow for model 1 without feedback} \\
ax_2 & \text{for model 2 with linear positive feedback}
\end{cases},
\tag{3.2}
$$

and kinetics

$$
v(u, \mathbf{x}, [PC], p_1, k_1) = v_1(u, x_1, [PC], p_1, k_1) - v_2(u, x_2, x_3, p_2, k_2)
\tag{3.3a}
$$

$$
v_1(u, x_1, [PC], p_1, k_1) = p_1 u \frac{x_1 [PC]}{x_1 [PC] + k_1}
\tag{3.3b}
$$

$$
v_2(u, x_2, x_3, p_2, k_2) = p_2 u \frac{x_2 x_3}{x_2 x_3 + k_2}.
\tag{3.3c}
$$

The state variables of our model are $\mathbf{x} = (x_1, x_2, x_3) = ([Cer],[DAG],[SM])$. All model variables are dimensionless and were normalized to the lipid levels of the endogenous system. Therefore, in the unperturbed system, steady state all variables have a value of 1. The input $u = [SMS1]$ describes the relative changes in the activity of SMS1, and was also normalized to the endogenous system, which equals a value of 1 in the control experiment. The model parameters d_1, d_2 and d_3 are rate constants of the linear degradation terms. The SMS reaction flux v was implemented via two Michaelis Menten kinetics v_1 and v_2. For the sake of simplicity we did not use the full substituted enzyme mechanism from chemical equation (1.5), but still wanted to account for possible substrate saturation effects. PC levels $[PC]$ were assumed to be constant and were permanently set to a value of 1, assuming a dominant external regulation. For the ceramide production C_{in} we assumed a constant inflow term $C_{in} = s_1$ for model 1, and a positive linear feedback $C_{in} = ax_2$ for model 2, depending on the amount of DAG.

3.3 Bayesian model analysis of the SMS1 models

To apply an *in silico* experiment to our model we needed to interpret the wet lab experiments. According to their experimental procedures, SMS1 overexpression experiments were achieved via stable transfections. They performed siRNA silencing of SMS1 for a period of over 60 hours, see Ding *et al.* (2008) for experimental details. Since altered SMS1 activities affected the cell cultures for a longer period of time we interpreted the data as new steady states of the reaction system. The respective inputs were set to $u_1 = 1$ for control, $u_2 = 2.2$ for overexpression and $u_3 = 0.77$ for silencing, and the system was simulated until a steady state was achieved. Values for the inputs were taken from measurements of SMS1 activity assays. We used the data and empirical estimates of the standard errors of Tables 3.1 and 3.2 to formulate a likelihood function based on a log-normal error model, see equation (2.5). We defined the system outputs as $y_1 = x_1$, $y_2 = x_2$ and $y_3 = x_3$, and respective data as $z_1 = [Cer]$, $z_2 = [SM]$ and $z_3 = [DAG]$. The system parameter vector was defined as $\tilde{\theta} = (s_1, p_1, p_2, d_1, d_2, d_3, k_1, k_2)^T$ for model 1 and $\tilde{\theta} = (a_1, p_1, p_2, d_1, d_2, d_3, k_1, k_2)^T$ for model 2 using the introduced parameter $\log_{10}$-transformation from Section 2.1. After some initial maximum likelihood estimation with a local optimizer we created a bounded $\log_{10}$-uniform prior and drew six million samples for each model with a standard adaptive Metropolis scheme. A subsample of five thousands parameter vectors was used to generate the Bayesian model fit and further prediction scenarios. For additional information about the MCMC sampling and the MATLAB R2011b implementation we refer to Appendix B and Weber *et al.* (2013).

3.4   Results model fitting

Figure 3.2 depicts the Bayesian model fit for both models. We draw the means and standard deviations of the measurements from Tables 3.1 and 3.2 together with the steady state predictions of the respective model outputs. It can clearly be seen that model 2 including the linear feedback agrees much better with the steady state data. Especially the increased ceramide levels (Figure 3.2, Symbol: '!') in the overexpression experiment cannot be matched at all by model 1. When examining the first column of sub-plots within Figure 3.2, we observe that model 1 aims for a compromise fit of the steady state ceramide levels in all three experiments. The linear feedback model on the other hand reproduces the ceramide levels very well. However, in Weber *et al.* (2013) we additionally provide a mathematical proof that model 1 cannot qualitatively mimic the observed steady state data, while model 2 is able to do so for certain classes of feedback functions. These functions cover our proposed linear feedback between DAG and ceramide. To validate our model and further demonstrate the application of Bayesian predictions we used our model to test further literature observations.

3.5   Simple linear feedback model agrees with further data of experimental manipulations of SM synthesis

We were aware that we cannot expect quantitative predictions from model 2 for further literature experiments that are measured in different cell lines, since it was calibrated with normalized data and was initially not designed for that purpose. However, we still aimed for a qualitative validation by predicting the tendency in the outcome of additional literature experiments. For this reason we looked at observations from Villani *et al.* (2008) in HeLa cells and predicted the outcome of an experiment series with our model 2. Villani *et al.* (2008) observed that treat-

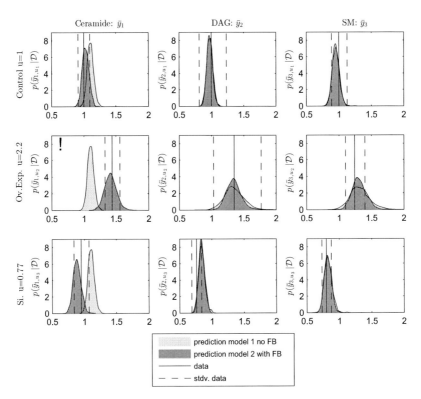

Figure 3.2: Bayesian data fit of models 1 and 2. The figure depicts the data from Tables 3.1 and 3.2 including standard deviations together with the predicted steady state distributions $p(\bar{y}_u|\mathcal{D})$ for the model variables ceramide, DAG and SM for the respective experiments. Sub-plots in row 1 correspond to the control experiment, row 2 corresponds to the overexpression and row 3 to the silencing experiments.

ment of HeLa cells with siRNA targeting SMS1 and addition of C_6-ceramide to the growth medium yields lower and higher PKD Golgi localization, respectively. Increased or decreased TGN localization was determined measuring overlay of signals of TGN and PKD markers in microscopy experiments. Since our feedback term represents accumulated biological effects, including PKD Golgi localization (see Figure 3.1), we used $C_{in} = ax_2$ as proxy for the amount of observed PKD localization at the TGN. This imposes that TGN-localized PKD is proportional to steady state DAG levels at the TGN. This assumption is very well motivated since PKD localization in presence of DAG has experimentally been observed multiple times, see Maeda *et al.* (2001), Baron and Malhotra (2002) and Table 1.1. The experimental findings of Villani *et al.* (2008) are summarized in Table 3.5 columns 1-4.

For validation we performed Bayesian predictions for the Villani *et al.* (2008) experiments by simulating the new scenarios for a representative parameter set of our calibrated model. For the case of SMS1 silencing we simulated the same scenario as for the respective experiment in the model calibration, but this time we plotted the feedback term. For the treatment of the cell culture medium with ceramide we added an additional artificial inflow term for ceramide to the system, assumed to be ten fold higher than the standard inflow rate C_{in} in the control experiment.

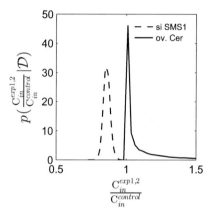

Figure 3.3: Predicted distribution of the normalized steady state model feedback term $C_{in}^{exp1,2}/C_{in}^{control}$ when simulating the experiments: ceramide overexpression and SMS1 silencing. Respectively, an amplification and damping of the feedback relative to the control (x-axis value 1) is observed.

Figure 3.3 depicts the distribution of the feedback term relative to the control given our training data $p(C_{in}^{exp1,2}/C_{in}^{control}|\mathcal{D})$ for the two experiments. The dashed curve represents the feedback after SMS1 silencing while the solid curve is the predicted outcome after addition of ceramide. We can see that the predictions are in qualitative agreement with the outcome of the experiments. In conclusion, we were able to perform a qualitative validation of model 2 using Bayesian model predictions. Final results are summarized in Table 3.5 columns 4-6.

Table 3.5: Summary of the findings of Villani *et al.* (2008) compared with our model predictions. In the first experiment siRNA is used to silence SMS1, in a second experiment cellular ceramide levels are increased. Both times PKD levels at the TGN are monitored via microscopy. Additionally, agreement between predictions and data is depicted.

Exp.	Source	Manipulation	Measured TGN-PKD	Prediction	Agreement
1	Fig. 12 A-C	SMS1 silencing ↓	Signal markers ↓	C_{in} ↓ Fig. 3.3	✓
2	Fig. 11 E-G	Ceramide addition ↑	Signal markers ↑	C_{in} ↑ Fig. 3.3	✓

3.6   Discussion: Model study supports the existence of feed- back regulations at the TGN

Summarizing the results of the model comparison study, we can draw several conclusions. We have seen that an isolated model of the SMS1 reaction was not able to reproduce lipid steady state measurements (Ding *et al.*, 2008). Furthermore, a model with a simple feedback extension agreed with the data set and correctly predicted the qualitative outcome of two further independent experiments (Villani *et al.*, 2008). We can surely argue that the data set itself has its weaknesses: It contains only information of average abundances within the cells and it is based on two different organisms (macrophages and CHO cells). The validation experiments are based on a third cell line (HeLa cells). So results are strongly dependent on the homology in the regulation of the SMS1 reaction in these organisms. A beneficial property of the experiments is that all interventions in the data selectively targeted SMS1, whose TGN localization is well identified (Huitema *et al.*, 2004; Tafesse *et al.*, 2007). This strongly increases the plausibility that the observed lipid level changes are a consequence of TGN related processes. Likewise, the manipulations in the validation experiments targeted SMS1 and ceramide, and monitored TGN specific changes via microscopy. Even though the simple model was not designed for quantitative predictions, results highly motivate further analysis of the feedback mechanism, which is yet modeled in an oversimplified way.

One could argue that the increased ceramide levels could also be a consequence of the retrograde hydrolysis of sphingomyelin at the plasma membrane or the lysosome (Hussain *et al.*, 2012). Alternatively it could involve the ceramide signaling pathway at the PM. In this pathway, SM is converted by acid SMase into ceramide in response to external stimulation (Hannun and Obeid, 2002). However, these pathways are not directly related to SMS1 perturbation and consume SM to generate ceramide, which was not observed in the data.

In contrast, the pathway involving PKD has the potential to directly trigger up-regulation of the de novo ceramide production by exhausting it at the ER via CERT mediated ceramide transfer mechanisms, see Figure 3.1 B.

Based on this motivating example, in the following chapter we will investigate the interplay of PKD, PI4KIIIβ and CERT in HEK293T cells with our own tailored wet lab experiments and Bayesian model analysis methods.

Chapter 4

Modeling interactions of the TGN key-players PKD, PI4KIIIβ and CERT

In this chapter we present the findings of a data-driven modeling study of the TGN key-players PKD, PI4KIIIβ and CERT. We are using a holistic model-based approach to face biological issues that are difficult to address with isolated experiments. We first formulate biological questions about the organization of the TGN and the regulation of ER to TGN ceramide transport. Then we introduce our wet lab experiments with HEK293T cell cultures to address the biological questions. Thereby we focus on different data types and introduce the experimental procedures needed to generate them. They comprise absolute measurements and time series data from dynamic perturbation experiments. We proceed with details on data post-processing and normalization methods. Based on the experimental data, we formulate and compare competing model hypotheses. For this purpose we employ ODE models based on chemical reaction kinetics. Like in Chapter 3, we follow the principle of data-driven modeling, choosing the model adequately complex to answer our major questions and compact enough to cope with the size of the data set. In contrast to Chapter 3, we here utilize our own data set, exclusively generated for the model study. Due to the more complex models with a larger amount of unknown parameters in this chapter, model calibration is more involved. We encounter this by applying Bayesian techniques for parameter inference such as parallel tempering MCMC algorithms. Furthermore, we test our model by simulating the outcome of new experimental scenarios that have not been considered for model calibration. Subsequently, we compare the simulations to the respective data of these wet lab experiments.

The model-based analysis provides major biological findings. Using Bayesian model comparison techniques the hypothesis of a cooperative behavior between PKD and CERT in the regulation of ceramide transfer between the ER and the TGN is supported. Specifically, our superior model features interrelated positive feedback loops including active PKD and ceramide transfer rates. Additionally, we identify active PKD to be the dominant regulator of CERT dependent ceramide transfer. Using Bayesian model predictions we are able to simulate biological scenarios that are difficult to access experimentally, like the interaction strengths and the operating point of the endogenous TGN.

Furthermore, we discuss our results by reviewing the state of the art literature concerning regulation mechanisms at the TGN and show the potential and limitations of our TGN-model. Finally, we close the Chapter by proposing conclusive model extensions that provide an additional level of detail for important future applications.

4.1 Contributions

The findings in this chapter are generated within a project based on a collaboration between the Institute of Systems Theory and Automatic Control (IST) of the University of Stuttgart and the Institute of Cell Biology and Immunology (IZI) of the University of Stuttgart.

All findings presented in Chapter 4 are based on the publication Weber *et al.* (2015). For an overview of the project in German language we refer to Weber *et al.* (2014).

All wet lab procedures were performed by Mariana Hornjik (IZI) including the generation of the raw image data.

Early models are described in the Diploma Thesis of Thomas Michael Hamm written at the IST (Hamm, 2014).

I established and maintained the models and updated them to the presented status. I chose the analysis methods, decided on their details, implemented them and performed the numerical analysis. I additionally performed all image data post-processing, data normalization and the absolute quantification scheme including error calculus.

4.2 Biological questions concerning the TGN

Regulation mechanisms at the TGN are not described yet by a quantitative model, hence there are many open questions to address. Keeping the biological background of Chapter 1 in mind we formulated the following questions:

1. Which CERT mediated ER-to-TGN ceramide transfer mechanism is supported by our data? (Background in Section 1.3.3)

2. What is the quantitative picture of the regulatory interactions at endogenous TGN? (Background in Section 1.3.5)

3. Which key-player regulates ER-to-TGN ceramide transfer at the MCS? (Background in Section 1.2.1)

Question one is of basic interest, since different biological hypotheses exist in the literature about mechanisms behind CERT mediated ceramide transfer between the ER and the TGN (see

Section 1.3.3). Understanding CERT transfer mechanisms, improves the biological understanding of processes at ER-TGN membrane contact sites, which are still poorly understood (Levine, 2007). Since the biological hypotheses translate into structurally different models, this question is addressed by model comparison. We furthermore study dynamic perturbations of the system since such data contain information about underlying regulation mechanisms.

Question two focuses on creating a quantitative biological model including key-player abundances and reaction fluxes. Up to now only qualitative biological interaction graphs exist such as Figure 1.3 in Section 1.3.5. However, a quantitative model allows for an insight into the endogenous network in unperturbed cells which is of high interest for basic research. To address this question we include absolute data into our study and perform quantitative simulations.

Question three addresses the identification of regulators for ER-TGN ceramide transfer. Studying the regulation of NLT of ceramide is important since this process is connected with secretion and vesicle budding, see Section 1.2.1, 1.2.2 and 1.2.3. These insights are also gained via simulation studies. Identification of regulators establishes the basis for future applications such as optimization of secretion rates (Florin et al., 2009) and allows for targeted interventions of the TGN regulatory network.

We now continue to introduce the experiments that are designed to address these questions in combination with mathematical modeling.

4.3   Results wet lab experiments

4.3.1   Overview about the experiments

In this section we explain the experimental procedures including cell culture experiments and data post-processing. All raw data is supplied in the electronic supplement of Weber et al. (2015). Sections 4.3, 4.4 and 4.5 require basic knowledge about cell culture handling, Western Blotting and image data quantification since explanations are kept on a medium level of detail to ensure compactness of this thesis. We start by explaining some initial decisions regarding the experiments, followed by the design of the wet lab experiments. Herein, we explain the results and details of the dynamic perturbation experiments. Then we move on to the image data post-processing and data normalization. For full detail of the used reagents and general cell culture handling we refer to the section 'Materials and Methods' in Weber et al. (2015) and references therein.

4.3.2   Using protein measurements from HEK293T cells

The data used for calibrating the models in Chapter 3 were based on measurements of average cellular lipid levels including ceramide, PC, SM and DAG. However, in this Chapter we do not consider any lipid measurements for reasons explained in the following.

Lipid concentrations can be determined on an average cellular level, see (Ding *et al.*, 2008). However, such measurements are lacking information about which particular sub-pool of lipids has changed. We already reflected this aspect in Section 3.6 and classified it as one of the weaknesses of the respective data set. We additionally addressed lipid localization in the biological introduction and highlighted that most TGN lipids are present in multiple membranes, especially DAG (Carrasco and Mérida, 2007) and ceramide (Hussain *et al.*, 2012). At the time of the study there was no experimental setup available to measure specific lipid sub-pools and we decided to solve the problem differently. Particularly, we used proteins as proxies for lipids and the overall TGN status and designed a protein focused model study. Regarding proteins, literature provides a lot of information about TGN localization and functional interactions, especially for PKD, PI4KIIIβ and CERT (Maeda *et al.*, 2001; Baron and Malhotra, 2002; Godi *et al.*, 1999; Haynes *et al.*, 2007; Hanada *et al.*, 2003). For this reason we focused on these three proteins in this study. The findings about protein localization are summarized in Table 2.1. Furthermore, most of these data are based on HEK293T cells which was the reason to also use this cell line for our model study.

4.3.3   Data and experiment types

Generally the experiments can be classified into two types, absolute quantification and time series measurements. In the absolute quantification experiments a multi-step procedure was performed to quantify proteins in units of average copy numbers per cell. This involved the establishment of standard curves, linear regression, several signal comparison steps and tracking of error propagation. These experiments form the basis for quantitative modeling and facilitate predictions of absolute abundances after model calibration. In our special case, these experiments require several replicates and multiple experimental steps to gain a small amount of absolute measurements. Therefore, this experimental protocol is only suited to measure the most important abundances instead of large time series of multiple proteins.

To gain insight into dynamic responses experiments that monitor the relative intensity changes of Western Blot signals of phospho-specific antibodies were performed. These data sets contain no absolute information but allow for the quantification of several time instants on a single Western Blot. The time series data was acquired in three steps. First, initial screenings with multiple perturbations and little time resolution were performed to scan for experiments that cause a significant system responses of multiple key-players. Then, we singled out the experiments that caused remarkable signal responses and they were repeated with a finer time resolution. In a last step, we used model-based experiment design to further improve remaining uncertainty in predicted scenarios by adding additional measurements to the data set.

4.3.4   Absolute data

The main challenge in absolute quantification of proteins is the lack of commercially available pre-quantified protein solutions. These pre-quantified solutions are required to establish standard curves that can be used for comparison with the signals of the samples. In our case, a standard solution was only available for PI4KIIIβ (GST-PI4KIIIβ, Vector Laboratories), but not for PKD and CERT. In order to compensate the lack of the other protein standards we decided to use the available PI4KIIIβ standard to evaluate our own quantification method, which is independent of protein specific standards. After the validation we subsequently applied this novel method to quantify the key-players PKD and CERT. In the next section, we explain how we quantified the protein PI4KIIIβ on an absolute scale without a specific standard.

4.3.4.1   Step 1: GFP standard and linear regression

First a commercially available GFP solution was used (recombinant purified GFP, Vector Laboratories) to generate four replicates of a dilution series of seven different concentrations. A representative replicate is depicted in Figure 4.1 A.

Signal intensities were quantified according to Section 4.4. No normalization was applied here to track Blot-to-Blot variability. Abundances μ_{MPL} in units of molecules per WB-lane served as dependent variables and signal intensities μ_{SPL} in arbitrary units as independent variables to calibrate a standard linear regression model in the form of

$$\mu_{\text{SPL}} = a\mu_{\text{MPL}} + b. \tag{4.1}$$

Here μ_{MPL} is the molecule number per lane, μ_{SPL} is the GFP-signal per lane, while a and b are maximum likelihood estimates of the parameters of the linear regression model. Additional fitting statistics such as parameter standard errors and R-values were directly obtained from the `MATLAB2011b` linear regression tool and listed in Figure 4.1 B. The usage of replicates from multiple gels allowed to use a total amount of $N = 28 \ (= 4 \times 7)$ data points for model calibration. Hence, the regression model accounts for Blot-to-Blot variability and can be used to quantify samples from new gels.

4.3.4.2   Step 2: Abundance estimation of ectopically expressed proteins

After the establishment of the GFP-standard, we proceeded with determining abundances of ectopically expressed proteins via the GFP standard. In three replicate experiments lysates from HEK293T cells were prepared that were ectopically expressing a GFP-tagged version of the target protein, here GFP-PI4KIIIβ. Along with the cells expressing GFP-PI4KIIIβ three wells of untreated cells were grown. The culture confluence was determined and all six wells were harvested after 24 h.

The three samples from the ectopically expressing cells were visualized via WB and GFP signals were quantified. For the image data of this step we refer to the electronic supplement of

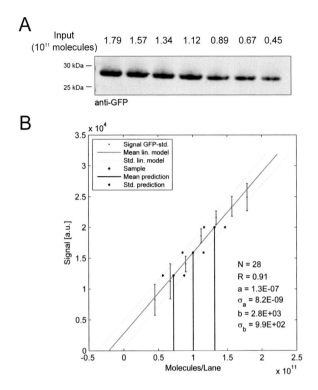

Figure 4.1: GFP signal standard curve. (A) A dilution series of recombinant GFP was loaded on a polyacrylamid gel and visualized by Western blot analysis using a GFP-specific antibody. (B) Results of a linear regression model analysis using the data from GFP signals.

Weber *et al.* (2015). Subsequently, the regression model from equation (4.1) was used to inversely predict the expected abundance of GFP-tagged target protein in the lanes, according to

$$\mu_{\text{OML}}^i = \frac{\mu_{\text{OSL}}^i - b}{a}. \qquad (4.2)$$

Here μ_{OML}^i represents the molecule number per lane of sample i for the ectopically expressed proteins, while μ_{OSL}^i are the corresponding GFP signals per lane. Inverse regression results are depicted as black vertical lines in Figure 4.1 B.

To track the uncertainty of the inverse prediction we applied the Delta Method (Parker *et al.*, 2010) to estimate the standard error according to

$$\sigma_{\text{OML}}^i = \frac{\sqrt{\frac{SS_{\text{res},y}}{N-2}}}{a} \cdot \sqrt{1 + \frac{1}{N} + \frac{(\mu_{\text{OML}}^i - \bar{x})^2}{\sum_{i=1}^{N}(x_i - \bar{x})^2}}. \qquad (4.3)$$

The calculation requires the sum of squares of the y-residuals $SS_{\text{res.y}}$ of the linear regression, the data volume N for calibrating the regression model, the mean $\bar{x}$ of the x-values x_i and the regression model estimate of ectopically expressed molecules per lane μ_{OML}^i, see equation (4.2). For further details regarding the calculation of these values we refer to Parker *et al.* (2010).

The signals were measured from samples with different lane loads in the WB. Altering lane loads is a standard wet lab procedure to screen for loading artifacts and decrease the chances of saturated signals. Therefore we first normalized all predicted lane means and standard errors to the lane loads according to: $\mu_{\text{OMC}}^i = \mu_{\text{OML}}^i / N_{\text{cells}}^i$ and $\sigma_{\text{OMC}}^i = \sigma_{\text{OML}}^i / N_{\text{cells}}^i$. After the normalization μ_{OMC}^i represents the average ectopically expressed molecules per cell and σ_{OMC}^i the respectively scaled standard errors. Theoretically, the means should all be the same after this normalization, but measurement noise of the WB method and the regression model prediction uncertainty demand for a weighted estimate of the mean and standard error at this step according to

$$\hat{\mu}_{\text{OMC}} = \frac{\sum_{i=1}^{N} \sigma_{\text{OMC}}^i \mu_{\text{OMC}}^i}{\sum_{i=1}^{N} \sigma_{\text{OMC}}^i}, \tag{4.4}$$

$$\hat{\sigma}_{\text{OMC}} = \sqrt{\frac{\sum_{i=1}^{N} \sigma_{\text{OMC}}^i (\mu_{\text{OMC}}^i - \mu_{\text{OMC}})^2}{\sum_{i=1}^{N} \sigma_{\text{OMC}}^i}}. \tag{4.5}$$

After applying the calculations above we obtained first estimates of the average protein per cell $\hat{\mu}_{\text{OMC}}$ and the standard error $\hat{\sigma}_{\text{OMC}}$ for the ectopically expressed proteins. These estimates serve as basis for the calculations of the endogenous protein abundances.

4.3.4.3 Step 3: Estimation of endogenous abundances

In the final step we determined the endogenous PI4KIIIβ abundance from the protein abundances in the ectopically expressing cells via signal comparison. Again Western Blots of three lysates of the previously mentioned ectopically expressing cells were processed together with three untreated cell culture lysates. In contrast to previous detections, the membranes were now incubated with a protein-specific and not tag-specific primary antibody, here anti-PI4KIIIβ), see Figure 4.2 A.

Remark: Indeed the samples of the endogenous cells and the samples of the ectopically expressing cells were processed already on the same gel for the procedures in Section 4.3.4.2. The second gel was omitted since a two-channel infrared-detection device was employed. Hence, incubation with multiple secondary antibodies, here anti-GFP and anti-PI4KIIIβ, allowed for multiplexing. Summarizing, anti-GFP signals were used for the quantification of the signals in Section 4.3.4.2, while anti-PI4KIIIβ signals were used for successive steps.

We expected that signals of endogenous and ectopically expressed proteins have rather large intensity differences. Hence, we adapted the molecule numbers in the samples beforehand by altering the dilution factor in the cell lysis step. It turned out that a difference in dilution by a factor of five, i.e. $200\mu L$ lysate volume for endogenous cells and $1000\mu L$ lysate volume for

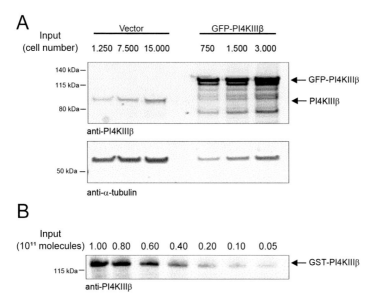

Figure 4.2: Signal comparison of endogenous and ectopically expressed proteins. (A) HEK293T cells were transfected with an empty vector as control or a plasmid encoding GFP-tagged PI4KIIIβ and incubated for 24 h. Cells were lysed, and signals of endogenous and ectopically expressed PI4KIIIβ were determined via Western blot analysis with a PI4KIIIβ-specific antibody. (B) A dilution series of recombinant GST-tagged PI4KIIIβ was loaded on a polyacrylamid gel and visualized by Western blot analysis using a PI4KIIIβ-specific antibody.

ectopic expression, provided adequate signals. As a direct consequence of this dilution step, ectopic expression signals y^i_{OSL} and endogenous signals y^i_{ESL} had to be normalized with their respective amount of cells per lane $N^i_{\mathrm{O.cells}}$ and $N^i_{\mathrm{E.cells}}$:

$$y^i_{\mathrm{ESC}} = y^i_{\mathrm{ESL}}/N^i_{\mathrm{E.cells}} \tag{4.6}$$

$$y^i_{\mathrm{OSC}} = y^i_{\mathrm{OSL}}/N^i_{\mathrm{O.cells}}. \tag{4.7}$$

Now that signals were comparable, we determined the individual signal ratios y^i_{rat}, the expected value μ_{rat} and the standard error σ_{rat} of the endogenous and ectopic expression samples. Calculations were performed according to

$$y^i_{\mathrm{rat}} = \frac{y^i_{\mathrm{ESC}}}{y^i_{\mathrm{OSC}}} \tag{4.8}$$

$$\mu_{\mathrm{rat}} = \frac{1}{N} \sum y^i_{\mathrm{rat}} \tag{4.9}$$

$$\sigma_{\mathrm{rat}} = \sqrt{\frac{1}{N-1} \sum y^i_{\mathrm{rat}} - \mu_{\mathrm{rat}}} \, . \tag{4.10}$$

Finally, Gaussian error propagation was used for calculation of the mean μ_{EMC} and standard error $\sigma_{End.}$ of endogenous protein abundances in units of protein per cell:

$$\mu_{EMC} = \hat{\mu}_{OMC} \cdot \mu_{rat.} \tag{4.11}$$

$$\sigma_{End.} = \sqrt{\left(\frac{\partial \mu_{EMC}}{\partial \mu_{OMC}} \sigma_{Ov.Exp.}\right)^2 + \left(\frac{\partial \mu_{EMC}}{\partial \mu_{rat.}} \sigma_{rat.}\right)^2}. \tag{4.12}$$

The estimates were finally embedded in the data set for the absolute abundances.

4.3.4.4   Validation

To validate our quantification method, we again created a standard curve via preparing a dilution series, analogue to procedures in Section 4.3.4.1. This time we employed a commercially available GST-PI4KIIIβ protein solution, see Figure 4.2 B. Linear regression was also performed analogue to the GFP-standard. We reused the already quantified anti-PI4KIIIβ signals from Figure 4.2 A, to directly determine cellular protein abundances using the GST-PI4KIIIβ based regression model. Final results of the PI4KIIIβ quantification are depicted in Figure 4.3 using blue color for the commercial standard and black color for our estimates.

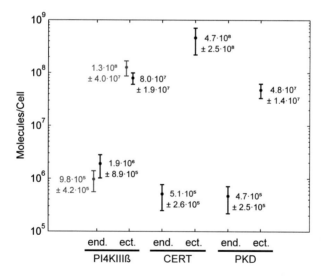

Figure 4.3: Absolute quantification of endogenoues proteins. Comparison of absolute quantifications via GST-PI4KIIIβ standard (blue) and GFP-standard (black).

Using our indirect GFP-method we obtained an estimate of $8.0 \cdot 10^7 \pm 1.9 \cdot 10^7$ PI4KIIIβ molecules per cell for the ectopic expression experiments, and $1.9 \cdot 10^6 \pm 8.9 \cdot 10^5$ molecules per cell for the endogenous protein amounts. Direct quantification with the commercial standard yielded estimates of $1.3 \cdot 10^8 \pm 4.0 \cdot 10^7$ and $9.8 \cdot 10^5 \pm 4.2 \cdot 10^5$ molecules per cell for the ectopic expression and the amount of endogenous protein, respectively. Mean estimates of both methods differ by a factor of less than two, and in particular, they agree in the order of magnitude. The Blot-to-Blot variability of WB methods combined with conservative error tracking methods resulted in final coefficients of variation between 25% and 40% in both methods. Summarizing these results we concluded that our approach provides reliable first estimates for the absolute abundance of PI4KIIIβ in both, the endogenous and the ectopically expressed scenario when compared to the commercial standard quantification.

4.3.4.5 Quantification of PKD and CERT.

To acquire additional absolute data we also applied the procedure for estimation of PKD and CERT abundances. Results are also shown in Figure 4.3. In the case of CERT, we have changed the tag of the expression construct to FLAG-CERT instead of GFP-CERT at a later stage. This was done to avoid occasionally observed interference with signals from other antibodies. Since FLAG-CERT runs on a different molecular weight signal crosstalk could be avoided in this way. However, endogenous CERT abundances were still determined via GFP-CERT constructs according to our method. In order to obtain estimates of FLAG-CERT abundances as well we employed an additional comparison step. We determined expression ratios of FLAG-CERT and calculated the ectopically expressed amount from the endogenous estimates as described in Section 4.3.4.3.

To account for the three biological replicates that were used to calculate the mean estimates of the absolute abundances in Figure 4.3, we also included these values as three identical data points in the final data set with their respective estimates for the standard errors. All raw image data for the quantification procedures are available in the electronic supplement of Weber *et al.* (2015).

4.3.5 Snapshot data

Some of the antibodies in the project have, to the best of our knowledge, been used for the first time for a quantitative modeling study and have an inadequate performance with respect to signal intensities when detecting endogenous protein abundances. This is especially true for the phospho-antibodies of CERT (α-phosphorylated serine 132) and PI4KIIIβ (α-phosphorylated serine 294), which are only suited for observations in ectopically expressed states. Solely the phospho-specific PKD1 antibody (α-phosphorylated serine 910) allows for endogenous detection. This has direct implications on experiment planning. In order to monitor time courses of the phosphorylation state changes of these proteins, we had to bring the cells in an ectopically expressing state beforehand.

After ectopic expression a PKD activator or inhibitor was added to the supernatant to trigger short-term responses. More precisely, activation of PKD1 was performed via potent activator phorbol 12,13-dibutyrate (PDBu, Sigma-Aldrich), see Rozengurt *et al.* (2005). Inhibition of PKD1 was performed via the highly selective inhibitor kb NB 142-70 (Tocris), see Bravo-Altamirano *et al.* (2011). Combining ectopic expression with activation and inhibition of PKD, we performed some initial snap shot experiments with our HEK293T cultures, see Figure 4.4. In the experiment in Figure 4.4 A, we ectopically expressed GFP-PI4KIIIβ for 24 h and subsequently applied a combination of inhibitor and activator treatments. We monitored changes in phospho-PKD and phospho-PI4KIIIβ signals and performed control measurements with anti-Tubulin and anti-GFP antibodies. From left to right, the lanes depict no additional treatment, activation with PDBu for 0.25 h, inhibition with kb NB 142-70 for 1h and a combined treatment of 1h inhibition and 0.25 h activation. In the experiment in Figure 4.4 B we ectopically expressed GFP-CERT for 24 h and applied the same subsequent manipulations as in the first experiment. This time we monitored the phosphorylation state of PKD and CERT while again running GFP and Tubulin specific controls. In the leftmost lanes of experiment A and B we can see that ectopic expression of PKD and PI4KIIIβ increases the phosphorylation signal of PKD. The effect was stronger for the CERT expression. In experiment A, phospho PKD showed more intensive signals after activation, decreased signals after inhibition and medium signals for the combined treatments. The phosphorylation signals of PI4KIIIβ behave in the same way. In experiment B phospho signals of PKD qualitatively behave in the same way. They again showed increased, decreased and intermediate signals, but with an overall stronger signal. Here, phosphorylated CERT showed no significant signal increase upon PKD activation, but had decreased and intermediate intensities for the inhibition and the combined treatment.

For planning the time series measurements we selected the manipulations that showed significant effects in the signals of both phosphospecific antibodies within experiments A and B. Hence, we choose the activator treatment in experiment A lane two, and inhibitor treatment in experment B lane three.

4.3.6   Time series data

We repeated the selected experiments and increased the number of measurements. Figure 4.5 shows the repetition of the two selected snapshot experiments, particularly ectopic expression of PI4KIIIβ with subsequent PDBu treatment and ectopic expression of CERT with subsequent kb NB 142-70 treatment. Again, the ectopic expression was performed for 24 h before the additional perturbations were applied. Subsequently, a total of six additional measurement times were chosen in both experiments: 0 h, 0.08 h, 0.16 h, 0.25 h, 0.5 h and 1 h for experiment A, and 0 h, 0.25 h, 0.5 h, 1 h, 2 h, 3 h for experiment B. For the inhibitor experiment, sampling times were scheduled up to three hours after the addition since decreased signals were observed on this time scale. The first lane of Figure 4.5 B shows signals prior to the ectopic expression. To avoid crosstalk, the vector expression construct for CERT was changed from GFP-CERT to FLAG-CERT, see Section 4.3.4.5 for the initial GFP-CERT signals. The outcome of both time series experiments met our expectations from the snapshot experiments in Figure 4.4.

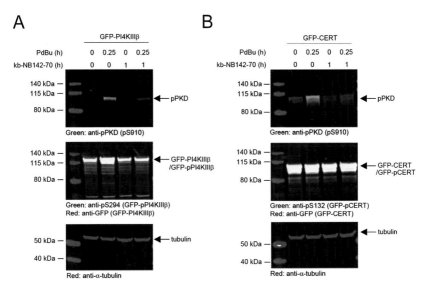

Figure 4.4: Western Blots of initial snapshot experiments. HEK293T cultures ectopically expressed GFP-PI4KIIIβ (A) and GFP-CERT (B) for $24h$. Subsequent single and combined treatments of the supernatant with PKD activator PdBu and PKD inhibitor kb NB 142-70 were performed. Phosphospecific antibodies for PKD, PI4KIIIβ and CERT monitor changes in the activation status of the proteins (green channel). Overall abundance of GFP-PI4KIIIβ and GFP-CERT was monitored with GFP-antibodies (red channel).

The PDBu treatment in experiment A caused a fast increase of the phospho-signals of PKD and PI4KIIβ within the next hour. In experiment B, ectopic CERT expression increased PKD phosphorylation. After addition of the inhibitor, both CERT and PKD phosphorylation levels decreased over the next three hours. All experiments were repeated at least three times. Some lanes with WB artifacts, see e.g. lane three in experiment B, were not used for quantification. In such cases additional experiments were performed, so that all measurement time points were represented with at least three replicates in the data set. Normalization of the time series signals was performed according to Section 4.5 and the data was added to the overall data set.

4.3.7   Refinement measurements

Guided by our model-dased experiment design method (see Weber *et al.* (2012)), we performed refinement measurements in the form of additional time series experiments.

The method is based on making Bayesian predictions as described in Section 2.9. In our case we used samples from the posterior distribution to predict the wet lab experiments that we already successfully performed. We predicted the outcome for additional measurement times of the measurable outputs and determined the variance in the predictions. Subsequently, we selected

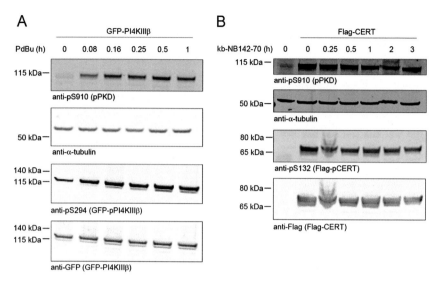

Figure 4.5: (A) HEK293T cells were transfected with a plasmid encoding GFP-tagged PI4KIIIβ and cultured for 24 h. Cells were stimulated with PDBu for the indicated time points, lysed, and phosphorylation and expression of PI4KIIIβ were analyzed by Western blot analysis using a pS294- and a GFP-specific antibody, respectively. Autophosphorylation of endogenous PKD was detected using the pS910-specific antibody. Detection of tubulin served as a loading control. Shown is a representative experiment, n = 3. (B) HEK293T cells were transfected with a control vector or a plasmid encoding Flag-tagged CERT and cultured for 24 h. Control cells were left untreated, whereas Flag-CERT transfected cells were treated with the PKD selective inhibitor kb NB 142-70 for the indicated time points. Afterwards cells were lysed, and phosphorylation and expression of CERT were analyzed by Western blot analysis using a pS132- and a Flag-specific antibody, respectively. Autophosphorylation of endogenous PKD was detected using the pS910-specific antibody. Detection of tubulin served as a loading control. Shown is a representative experiment, n = 3.

the outputs and time points with the largest variances as future candidate measurements. After feedback with the experimentalists we excluded some inconvenient candidate measurements, such as measurements scheduled outside time frames of typical cell culture treatments. The experiment design suggested to measure signals of phospho PKD in experiment A at 3 h and 6 h directly after the ectopic expression of PI4KIIIβ is initiated. Additionally measurements of phospho-PKD signals 3 h and 6 h after the PDBu activation were suggested. We performed these experiments and included them into our data set. We discuss the results of these experiments in Section 4.6.5.

4.4   Image data post-processing and optical densitometry

All Western Blots were scanned using the `Li-Cor ODYSSEY` infrared imaging system, based on the scanning software version 3.0.16. Membranes were scanned using the standard channel hardware gain settings with a value of 5 for both the $700nm$ and $800nm$ excitation. Images were checked for saturation effects in protein bands that were used for further quantification (saturated areas are indicated by the software). After the scan the software gain function was used to adjust the intensity of the image. This software gain was adjusted separately for each channel between an intensity level of 3 to 8, depending on the antibody in use. Afterwards the softwares *.tif format export functions were used to export gray-scale images for each channel. Gain information was added to the exported image data. All *.tif image data was further processed using a `MATLAB R2011b` script based on the software's image analysis toolbox. Here, the gray-scale '*.tif'-images were imported into `MATLAB R2011b` as image-matrix with intensity values ranging from 0 to 255.

We used our own `MATLAB R2011b` quantification script that is based on the findings of Gassmann *et al.* (2009) in the field of WB-analysis and optical densitometry. Here, we summarize the main features of the script and show an image quantification example. The following image post-processing features were added in our software (compare Table 1, Gassmann *et al.* (2009)):

- Integral of the optical density profile is used as final value.

- Background correction is calculated for individual lanes, using baseline subtraction.

- Width of the sample tool is defined as one third of the band width and centered to the middle of the band.

According to Gassmann *et al.* (2009), the combination of a local background correction with the use of the optical density profile as final scalar value, yielded the best results (R-values) when establishing a standard curve with samples with known abundances. The usage of only one third of the protein band width as sample tool geometry is especially useful in cases were antibodies tend to have artifacts on the edges of the protein bands. These effects were indeed observed in our signals (see Figure 4.6 B), so we added this feature to the software. As a rule of thumb, these artifacts are more frequently observed when proteins of different molecular masses are quantified on the same gel.

An example of a post-processing procedure of a scanned Western Blot image is depicted in Figure 4.6. In the example a quantification of a signal from an anti phospho-PI4KIIIβ antibody (α-PS294) was performed.

After the scanning process, information about intensities of all channels, including saturated areas, is available in the `Li-Cor ODYSSEY` software internal project formate (Figure 4.6 A). Single or multiple channels can be exported as '*.tif'-images after adjusting the software gain and clipping image parts (Figure 4.6 B). In this particular case single channel export was chosen and a gray-scale '*.tif'-image was generated. In the case of multi-channel export, a RGB color '*.tif'-image is created and information of the $800nm$ channel is stored in the red values and the $700nm$ channel is stored in the green values of the image. Saturated areas are stored in the blue values of the image and appear as white or blue areas, for single or multichannel export respectively. Figure 4.6 B depicts a snapshot while running our `MATLAB R2011b` image analysis script. The user is asked to communicate the software the lanes desired for quantification. For the selected protein bands, one third of the average lane width was calculated and the sample tool was centered automatically to each individual Western Blot band. Figure 4.6 B additionally displays red circles where artifacts appear in the WB (manually added to the snapshot for demonstration purposes). Figure 4.6 C depicts the calculations of the automatic background subtraction for the two previously selected lanes in Figure 4.6 B. After confirmation of the background subtraction, the software calculates the integrals of the optical profile and stores the results as `MATLAB` variables, see Figure 4.6 D.

4.5   Data normalization

After the signals of the bands for the individual antibodies were available as scalar values from the image data quantification, they were post-processed via normalizations. The data normalization was performed for different purposes depending on the respective experiment type:

- to account for loading differences in the WB process

- to reduce Blot-to-Blot variability

- to calculate ratios or relative changes in signals

In this subsection we introduce the normalization procedures especially applied to the time series data before they were used for model calibration. For time series data, typically the changes in the signals of the phosphorylation states of the respective proteins are monitored. Here, these proteins are PKD, PI4KIIIβ and CERT. In our case, anti-phospho-PKD (α-pS910) is the only phospho-specific antibody employed, that is suited for detection of endogenous protein singnals. For reasons to be explained, we applied a different normalization to phospho-PKD signals than for phospho-PI4KIIIβ and phospho-CERT signals, since these two proteins were studied in ectopically expressed states.

4.5.1   Normalization of anti-phospho-PKD signals

Regarding phospho-PKD signals, we normalized signals of α-pS910 antibody to the loading control signal of the anti-tubulin antibody of the same lane on the gel. This has the purpose of reducing the influence of loading differences from wet lab WB preparation. Assuming that all signals are detected in the linear range, this normalization provides a signal $z_{PKD,norm.tub.}$, which

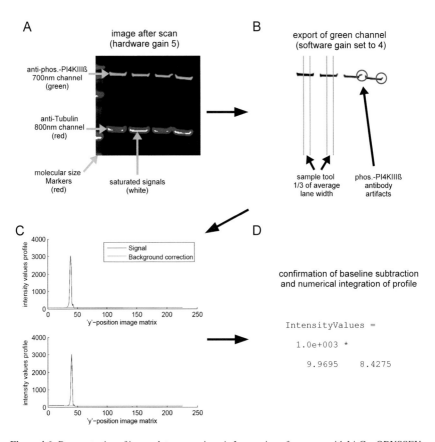

Figure 4.6: Demonstration of image data processing. A: Image view after a scan with Li-Cor ODYSSEY software. Signals of two antibodies (anti-tubuling and anti-phospho-PI4KIIIβ) are depicted in red (800nm channel) and green (700nm channel). Saturated areas are dyed in white. B: Exported gray-scale image of the green channel. Red circles highlight antibody specific artifacts, green lines depict the sample tool width. C: Background correction based on baseline subtraction, performed individually for each lane. D: Final values of the integral of the optical density profile.

is proportional to the amount of phosphorylated PKD $x_{\text{phospho. PKD}}$, according to

$$z_{\text{PKD,norm.tub.}} = \frac{z_{\text{p-PKD}}}{z_{\text{Tubulin}}} \propto x_{\text{phospho. PKD}},\qquad(4.13)$$

where $z_{\text{p-PKD}}$ and z_{Tubulin} are the signals of the α-pS910 and the anti-tubulin antibody, respectively.

4.5.2   Normalization of anti-phospho-PI4KIIIβ and anti-phospho-CERT signals

In the case of phospho-PI4KIIIβ and phospho-CERT signals we applied a different normalization procedure. These proteins were measured in ectopically expressed states since the α-pS293 and α-pS132 antibodies signals are not suited to measure the phosphorylation status under endogenous conditions. Ectopic expression experiments amplify the signal of these antibodies since they increase the overall amount of the target protein in the cells. More precisely, these experiments add large amounts of a tagged version of the same protein to the cell (GFP-PI4KIIIβ and FLAG-CERT). As a consequence, the overall cellular abundance becomes dominated by the artificially introduced tagged protein. Thereby, an ectopically expressed protein can reach abundances up to a thousand times higher than the natural equivalent, see results in Figure 4.3 on page 54. Since the abundance of the endogenous isoform can be neglected, we concluded that the overall protein abundance is approximately proportional to the signal of the antibody that binds the respective tag of the expression construct, here α-FLAG or α-GFP.

Anti-tag antibodies are generally very well characterized and we used their signals for normalization. However, other than the tubulin normalization, the normalization to the tag-signal results in different signal characteristics. The normalized signals $z_{\text{PI4KIIIβ norm.tag.}}$ and $z_{\text{CERT norm.tag.}}$ are proportional to the phosphorylated fraction of the overall protein according to

$$z_{\text{PI4KIIIβ norm.tag.}} = \frac{z_{\text{p-PI4KIIIβ}}}{z_{\text{anti-tag}}} \propto \frac{x_{\text{PI4KIIIβ phospho. prot.}}}{x_{\text{PI4KIIIβ total. prot.}}} \tag{4.14}$$

$$z_{\text{CERT norm.tag.}} = \frac{z_{\text{p-CERT}}}{z_{\text{anti-tag}}} \propto \frac{x_{\text{CERT phospho. prot.}}}{x_{\text{CERT total. prot.}}}. \tag{4.15}$$

Here, in the case of CERT, $z_{\text{p-CERT}}$ is the signal of the respective phospho-antibody, $z_{\text{anti-tag}}$ is the signal of the respective tag-antibody, $x_{\text{CERT phospho. prot.}}$ is the absolute amount of phospho-protein and $x_{\text{total. prot.}}$ is the absolute amount of total protein (analogously for PI4KIIIβ). This normalization gives us a signal containing information about the relative phosphorylation grade of PI4KIIIβ and CERT, which is an important information, especially when combined with absolute measurements of the respective proteins. Needless to say signals proportional to the relative phosphorylation state are inherently decoupled from loading differences, so we did not need any additional tubulin normalizations.

4.5.3   Additional normalizations at predefined time instants

Despite enormous efforts to improve the precision of wet lab procedures, WB signals still have a Blot-to-Blot variability for replicate experiments with a coefficient of variation of about $c_v = 0.2$. Blot specific normalizations can help to reduce extreme Blot-to-Blot variability. Especially for the time series signals of our phosphorylation states we applied a subsequent Blot specific normalization. We chose a time point specific normalization, which was lately compared to other Blot specific normalization types by Degasperi *et al.* (2014). Among the tested normalization methods, this method generated the smallest amount of false positive results in their hypothesis testing scheme. This renders the method a conservative and reliable normalization procedure that reduces technical noise across replicate experiments and at the same time does not overly reduce the biological variance in the data.

Like the authors recommendations we always used a reliable signal within a specific time series experiment for this normalization. This means that we did not choose an overly weak signal (high signal to noise ratio) and no exceptionally strong signal (potentially saturated). Once defined, the same time point was then used for normalization of all replicates of this experiment. We applied the Blot specific normalization after the normalizations of Sections 4.5.1 and 4.5.2. For clarity we show a small example:

If a Blot specific normalization at $t = 24h$ is applied to a phospho-CERT signal, the twofold normalized signal reads as

$$z_{\text{p-CERT, 24h}}(t) = \frac{z_{\text{CERT norm.tag.}}(t)}{z_{\text{CERT norm.tag.}}(t = 24h)} = \frac{\frac{z_{\text{p-CERT}}(t)}{z_{\text{anti-tag}}(t)}}{\frac{z_{\text{p-CERT}}(t=24h)}{z_{\text{anti-tag}}(t=24h)}}. \tag{4.16}$$

The respective reference data point of the time instant used for normalization (in this example $t = 24h$) was removed from the data set, see also Degasperi *et al.* (2014).

After the introduction of the wet lab experiments, the image data post-processing and the data normalization we continue to the Bayesian model analysis.

4.6 Results Bayesian model analysis

In Section 1.3 we have introduced the biological background of the TGN key-players and in Section 4.3 we explained the data set and experimental options. We now explain how we derived our ODE model structures to answer the biological questions formulated in Section 4.2. Subsequently, we apply our advanced MCMC sampling methods to calibrate our models to the experimental data and compare the different model hypothesis. We validate our model and use Bayesian predictions to address our initial questions and discuss the results.

4.6.1 Modeling of TGN interactions

To answer the questions formulated in Section 4.2 we first established two simplified models in the form of biological cartoons, see Figure 4.7. A main characteristic about the models is that they use the TGN key-players PKD, PI4KIIIβ and CERT as proxies for the TGN status, for reasons explained in Section 4.3.2. We considered the respective interactions between these players and illustrate them at the membranes of the ER and the TGN, according to the literature in Table 1.3 on page 20.

The details of the mechanism of CERT mediated ceramide transfer from the ER to the TGN are currently still unclear (Perry and Ridgway, 2005; Hanada, 2006, 2010). As described in

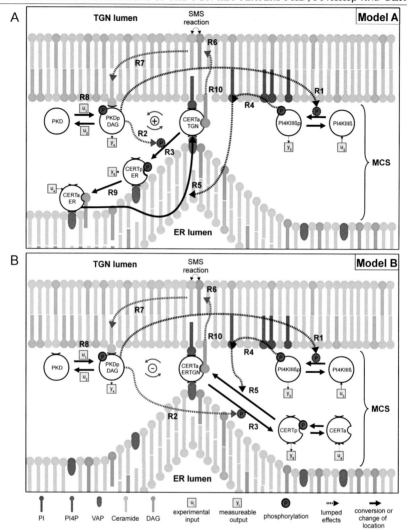

Figure 4.7: TGN interactions of PKD, PI4KIIIβ and CERT. PKD and PI4KIIIβ are modeled as inactive or active TGN bound forms (variables PKD, PKDpDAG, PI4KIIIβ and PI4KIIIβp, respectively). Active PKD activates PI4KIIIβ (R1) and detaches CERT from the TGN membrane (R3). Active PI4KIIIβ attracts CERT to the TGN (R5) by producing PI4P (R4). Models differ in CERT dependent ceramide transport: (A) Ceramide transfer is realized by a circular reaction scheme of CERT (R3, R9, R5). PKD induced CERT phosphorylation (R2) and ceramide dependent PKD activation (R10, R6, R7, R8) constitute a positive feedback mechanism (blue arrows). (B) CERT is simultaneously bound to the ER and the TGN (CERTaERTGN) to transfer ceramide (R10). Phosphorylation of CERT (CERTp, R5) causes an interruption of the transfer process and forms a negative feedback between PKD and CERT (blue arrows).

Section 1.3.3 two models were studied in literature (see also Figure 6 in (Hanada, 2010)). The 'short distance shuttle' model, which suggests that CERT travels through the cytosol, shuttling between ER and TGN membranes during the ceramide transfer process (Hanada *et al.*, 2007) and the 'neck-swinging' model that suggests that CERT 'shovels' ceramide from the ER to the TGN while CERT is simultaneously bound to both membranes (Hanada *et al.*, 2003; Kawano *et al.*, 2006). By studying the regulatory differences of the two literature transport mechanisms, we built two models A and B. In the two models PKD-mediated phosphorylation affects CERT transport activity differently. We identified that in the 'short distance shuttle' theory (model A) it has a positive effect on CERT mediated ceramide transfer. However, in the 'neck-swinging' theory (model B) it decreases transport activity. This generates a positive feedback loop in model A and a negative feedback loop in model B, between the ceramide transfer rate and PKD activity, as indicated by the blue edges in Figures 4.7 A and B. Furthermore, in the 'short distance shuttle' model CERT is part of a cyclic reaction scheme, which is driven by active PKD and active PI4KIIIβ. In the 'neck-swinging' model, CERT is part of a switch like reaction scheme. Here, CERT mediated ceramide transfer is switched on by active PI4KIIIβ and switched off by active PKD. Summarizing, models A and B display the regulatory differences of theories A and B, described in Section 1.3.3.

Solid arrows in Figure 4.7 represent chemical conversions such as phosphorylation or dephosphorylation reactions. Solid arrows can also imply bindings and detachment from membranes or a change of location. Dashed arrows depict lumped effects or catalytic influences. To emphasize that this cartoon is a result of a data-driven modeling approach we also depicted possible experimental inputs u. Additionally, we denoted measurable phosphorylation states with variables y, as described in Section 4.3.6. The choice of using a total of seven variables in both models is adjusted to be close to the count of six different measured signals, namely three phospho-states and three absolute quantities. We now explain the respective simplifications for each variable, which were made to adjust the model complexity to the amount of available data.

Concerning PKD and PI4KIIIβ the models are designed in a similar way. The kinase PKD is represented in the model in only two forms: We chose an inactive cytosolic form 'PKD', that is related to the TGN-ER MCS. Additionally, we included a phosphorylated and active representation that is bound to DAG 'PKDpDAG'. We assumed that all DAG bound PKD is instantly activated by PKCη (R8), which we did not include into the model due to lack of respective measurements. In the model, PKD is experimentally accessible via two interventions, inhibition and activation denoted by u_1 and u_2. Manipulating these inputs changes the balance between the active and inactive representation. We can measure active PKD via a phosphospecific antibody (y_4). For clarity the absolute output is not depicted but corresponds to the sum of the all PKD representations within the model. Active PKD triggers two regulatory effects: It causes phosphorylation of TGN related PI4KIIIβ (R1) and thereby activates the kinase in this process. Additionally, it colocalizes with TGN bound CERT, denoted by 'CERTaTGN' (or respectively 'CERTaERTGN' in model B) and detaches it from the TGN by initiating a series of successive phosphorylation reactions (R2).

The kinase PI4KIIIβ is also modeled using two representations. The phosphorylated active form 'PI4KIIIβp' and the unphosphorylated inactive form 'PI4KIIIβ' are both related to the near cytosol of the ER-TGN MCS. Active PI4KIIIβ attracts CERT towards the TGN membrane by initiating a reaction chain. It colocalizes with PI at the TGN membrane and phosphorylates it, thereby creating PI4P (R4). Subsequently, increasing amounts of PI4P lipids attract more CERT by binding its PH domain (R5). We experimentally access PI4KIIIβ by performing ectopic expressions, the respective input is denoted by 'u_3'. This experiment generates an inflow of additional GFP-PI4KIIIβ, thereby increasing the overall PI4KIIIβ abundance. Ectopically expressed GFP-PI4KIIIβ is treated equivalent to the endogenous protein in the model. Also here we can measure phosphorylated 'PI4KIIIβp' (y_5) and the absolute abundance.

We now focus on the differences in the CERT reaction scheme. Model A features three CERT representations: A TGN and ER bound form 'CERTaTGN', a phophorylated ER bound form 'CERTpER' and a dephosphorylated ER bound form 'CERTaER'. Here, 'CERTaTGN' is phosphorylated by PKD (R2) and detached from the TGN (R3). Subsequently, 'CERTpER' is dephosphorylated and extracts ceramide from the ER membrane (R9). In a further step, 'CERTpER' is attracted by PI4P and again attaches to the TGN (R5) where it donates ceramide to the TGN membrane (R10). The ceramide delivered to the TGN causes a feedback upon PKD activation (R7) by promoting the production of DAG over the SMS reaction (R6). The strength of this ceramide related feedback upon PKD is thereby dependent on the reaction rate of the circular reaction scheme of CERT.

Model B has an unphosphorylated, TGN and ER bound CERT representation 'CERTaERTGN' which is able to transfer ceramide from the ER to the TGN (R10) without the circular reaction scheme from model A, by the 'neck-swinging' mechanism. It also contains two further CERT representations. 'CERTp' is phosphorylated, transfer-inactive and related to the near cytosol at the MCS. 'CERTa' is unphosphorylated, transfer-inactive and also related to the near cytosol at the MCS. Here, CERT phosphorylation (R3) inhibits ceramide transfer by displacing 'CERTaERTGN' from its double bound status. The ceramide transfer rate (R10) in this model is dependent on the amount of double bound 'CERTaERTGN'. Mechanisms of CERT attraction via PI4KIIIβ (R5) and detachment via PKD (R3) are realized in the same way as in model A. Likewise, the model also features a feedback of ceramide transfer upon PKD activity via the SMS1 reaction.

Both models allow to manipulate CERT via ectopic expressions of FLAG-CERT or GFP-CERT via input u_4. In both models, we are able to measure phosphorylated CERT and the total abundance, corresponding to the sum of the respective three CERT representations. We now describe how to derive ODEs from the biological cartoons.

4.6.2   Reaction kinetics

Table 4.1 depicts the reaction kinetics we have chosen to represent the effects introduced in the previous section in Figure 4.7. The reactions are classified into three different types.

- Reactions 'S' define system inflows. They represent basal synthesis rates or artificial inflows from experimental interventions such as ectopic expressions.

- Reactions 'V' define conversions or change of locations. Total mass is preserved in this reaction.

- Reactions 'D' describe degradation rates. Degradation rates summarize all biological effects that reduce a species abundance in the model. This includes protein auto-degradation and reactions consuming the respective proteins that are not included in the model.

Different kinetics were chosen for the individual reactions with the aim to keep the overall model as simple as possible. All basal inflows 'S' are modeled as constants. Only variables representing unphosphorylated and unbound versions of PKD, PI4KIIIβ and CERT are modeled with a synthesis rate. We assumed that all basal and experimental system inflows enter our reaction system via these inflows. For basic conversion reactions linear mass action kinetics of the type

$$f_1(x_a, \tilde{\theta}_1) = \tilde{\theta}_1 x_a \qquad (4.17)$$

are used. In Table 4.1 this involves basal PKD phosphorylation (V_{12}) and dephosphorylation (V_{13}), PI4KIIIβ dephosphorylation (V_{21}) as well as all degradation rates 'D' in both models. Reaction equations for all variables in the model have these degradation rates. Linear kinetics are also used for CERT dephosphorylation at the ER in model A (V_{32}) and unspecific CERT de-/phosphorylation (V_{31}, V_{32}) in model B.

For regulatory interactions and lumped effects we used non-linear Michaelis Menten kinetics

$$f_2(x_a, x_b, \tilde{\theta}_1, \tilde{\theta}_2) = \tilde{\theta}_1 x_a \frac{x_b}{x_b + \tilde{\theta}_2} \qquad (4.18)$$

for both models.

Notably, we did not use Michaelis Menten kinetics in the classical sense but for lumped effects. We used the kinetic because the function f_2 has a saturation effect with respect to reactant x_b and it has only two parameters, a saturation threshold $\tilde{\theta}_2$ and a maximal reaction rate $\tilde{\theta}_1$. From a modelers perspective, including kinetics with saturation makes a lot of sense, if lumped effects cover complex biological events such as colocalizations at membranes. If multiple such effects are summarized with one reaction equation, it is very likely that at least one of these effects may have some limiting factors. The following reactions are modeled with f_2 type kinetics:

- PI4KIIIβ activation via PKD (V_{22} in both models)

- PKD activation via CERT dependent ceramide transfer (V_{11} in both models)

- Both models: CERT recruitment to the TGN membrane via PI4KIIIβ (V_{31} in model A, V_{33} in model B)

- CERT phosphorylation and TGN detachment via PKD (V_{33} in model A, V_{34} in model B).

To model the different ceramide transfer mechanisms, the feedback upon PKD activation (V_{11}) is implemented differently in models A and B. In model A, reaction rate V_{11} is dependent on the reaction rate V_{31} which represents the circular reaction scheme of CERT between the ER and the TGN. In model B, reaction rate V_{11} is dependent on the amount of double bound, transport active CERT x_7.

For all detailed information about the overall ODE model equations we refer to Table 4.1. It contains information about the ODE model structure, equations of the kinetics and model outputs and their respective normalizations (according to Section 4.5).

Table 4.1: Equations for ODE models A and B.
1. Column: Model structure and variables.
2. Column: Reaction rate equations.
3. Column: Output variables.

ODE structure	Reaction rate equations	Output variables
Subsystem 1: PKD		
$\dot{x}_1 = -V_{11} - V_{12} + V_{13} + S_{12} - D_{13}$	$V_{12} = f_1(x_1, \bar{p}_{12})(1 + u_5 \bar{p}_{45})$ → basal and experimental PKD activation	$y_1 = \dfrac{x_2(t)}{x_2(t=0h)}$
$\dot{x}_2 = +V_{11} + V_{12} - V_{13} - D_{14}$	$V_{13} = f_1(x_2, \bar{p}_{13})(1 + u_6 \bar{p}_{46})$ → basal and experimental PKD inhibition	$x_2(t=0h) \widehat{=}$ yPKDpN0
Subsystem 2: PI4KIIIβ	$S_{12} = \bar{s}_{12}$ → basal PKD production rate	$y_2 = \dfrac{x_2(t)}{x_2(t=24h)}$
$\dot{x}_3 = +V_{21} - V_{22} + S_{22} - D_{21}$	$D_{11} = f_1(x_1, \bar{a}_{11})$ → PKD degradation rate	$x_2(t=24h) \widehat{=}$ yPKDpN24
$\dot{x}_4 = -V_{21} + V_{22} - D_{22}$	$D_{12} = f_1(x_2, \bar{a}_{12})$ → active PKD degradation rate	$y_3 = \dfrac{x_2(t)}{x_2(t=25h)}$
Variable names:	$V_{21} = f_1(x_4, \bar{p}_{21})$ → deactivation of PI4KIIIβ	$x_2(t=25h) \widehat{=}$ yPKDpN25
$x_1 \widehat{=}$ PKD	$V_{22} = f_2(x_3, x_2, \bar{p}_{22}, \bar{m}_{22})$ → PKD mediated activation of PI4KIIIβ	$y_4 = \dfrac{x_4(t)/y_T(t)}{x_4(t=24h)/y_T(t=24h)}$
$x_2 \widehat{=}$ PKDpDAG	$S_{21} = \bar{s}_{21}$ → basal PI4KIIIβ production rate	$\widehat{=}$ yPI4K3BpRN24
$x_3 \widehat{=}$ PI4K3B	$S_{22} = \bar{p}_{u_3} u_3$ → experimental PI4KIIIβ production rate	$y_5 = x_1 + x_2 \widehat{=}$ yPKDt
$x_4 \widehat{=}$ PI4K3Bp	$D_{21} = f_1(x_3, \bar{a}_{21})$ → PI4KIIIβ degradation rate	$y_6 = x_3 + x_4 \widehat{=}$ yPI4K3Bt
	$D_{22} = f_1(x_4, \bar{a}_{22})$ → active PI4KIIIβ degradation rate	
Subsystem 3a: Model A	$V_{11} = f_2(x_1, V_{31}, \bar{p}_{11}, \bar{m}_{11})$ → PKD activation via CERT	
$\dot{x}_5 = -V_{31} + V_{32} + S_{31} + S_{32} - D_{31}$	$V_{31} = f_2(x_5, x_4, \bar{p}_{31}, \bar{m}_{31})$ → PI4KIIIβ dependent recruitment of CERT to TGN	$y_7 = \dfrac{x_6(t)/y_8(t)}{x_6(t=24h)/y_8(t=24h)}$
$\dot{x}_6 = -V_{32} + V_{33} - D_{32}$	$V_{32} = f_1(x_6, \bar{p}_{32})$ → dephosphorylation of CERT at the ER	
$\dot{x}_7 = +V_{31} - V_{33} - D_{33}$	$V_{33} = f_2(x_7, x_2, \bar{p}_{33}, \bar{m}_{33})$ → phosphorylation of CERT by PKD	$y_8 = x_5 + x_6 + x_7$
Variable names:	$S_{31} = \bar{s}_{31}$ → basal CERT production rate	$y_7 \widehat{=}$ yCERTpRN24
$x_5 \widehat{=}$ CERTaER	$S_{32} = \bar{p}_{u_4} u_4$ → experimental CERT inflow	$y_8 \widehat{=}$ yCERTt
$x_6 \widehat{=}$ CERTpER	$D_{31} = f_1(x_5, \bar{a}_{31})$ → degradation of active CERT at the ER	
$x_7 \widehat{=}$ CERTaTGN	$D_{32} = f_1(x_6, \bar{a}_{32})$ → degradation of inactive CERT	
	$D_{33} = f_1(x_7, \bar{a}_{33})$ → degradation of ER-TGN bound CERT	
Subsystem 3b: Model B	$V_{11} = f_2(x_1, x_7, \bar{p}_{11}, \bar{m}_{31})$ → PKD activation via CERT	
$\dot{x}_5 = -V_{31} + V_{32} + S_{31} + S_{32} - D_{31}$	$V_{31} = f_1(x_5, \bar{p}_{31})$ → CERT phosphorylation and binding to ER	$y_7 = \dfrac{x_6(t)/y_8(t)}{x_6(t=24h)/y_8(t=24h)}$
$\dot{x}_6 = +V_{31} - V_{32} - V_{33} + V_{34} - D_{32}$	$V_{32} = f_1(x_6, \bar{p}_{32})$ → CERT dephosphorylation and detachment from the ER	
$\dot{x}_7 = +V_{33} - V_{34} - D_{33}$	$V_{33} = f_2(x_6, x_4, \bar{p}_{33}, \bar{m}_{31})$ → PI4KIIIβ triggered CERT ER binding/dephos.	$y_8 = x_5 + x_6 + x_7$
Variable names:	$V_{34} = f_2(x_7, x_2, \bar{p}_{34}, \bar{m}_{33})$ → phosphorylation of CERT by PKD	$y_7 \widehat{=}$ yCERTpRN24
$x_5 \widehat{=}$ CERTa	$S_{31} = \bar{s}_{31}$ → basal CERT production rate	$y_8 \widehat{=}$ yCERTt
$x_6 \widehat{=}$ CERTp	$S_{32} = \bar{p}_{u_4} u_4$ → experimental CERT inflow	
$x_7 \widehat{=}$ CERTaERTGN	$D_{31} = f_1(x_5, \bar{a}_{31})$ → degradation of active CERT	
	$D_{32} = f_1(x_6, \bar{a}_{32})$ → degradation of inactive CERT	
	$D_{33} = f_1(x_7, \bar{a}_{33})$ → degradation of TGN-ER bound CERT	

4.6.3   ODE model parameters, inputs and outputs

The two models as stated in Table 4.1 both comprise seven state variables. The state variables are measured in units of molecules and cell, while reaction rates are defined as molecules per hour and cell. Since all reactions are assumed to be located at the near cytosol of ER-TGN MCS we did not define compartments. Models A and B have 26 and 27 unknown parameters, respectively. Parameters are arranged in groups: parameters $\tilde{p}_i$ are used for linear kinetics and maximum reaction rates in functions f_1 and f_2. Parameters $\tilde{m}_i$ denote saturation thresholds within kinetics of type f_2. Parameters $\tilde{a}_i$ define linear degradation rates. Parameters $\tilde{s}_i$ define basal synthesis rates of proteins. Parameters $\tilde{p}_{u_i}$ represent scaling parameters of system inputs. An effective $\log_{10}$-transformed parameter vector $\tilde{\theta}$ is implemented in the model such that $\tilde{\theta} = 10^{diag(\theta)}$ holds, and respectively for the single parameters $\tilde{p} = 10^p$, $\tilde{m} = 10^m$, $\tilde{s} = 10^s$, $\tilde{a} = 10^a$ and $\tilde{p}_u = 10^{p_u}$ holds.

For parameter estimation, the model parameter vector θ also contains the input scaling parameters denoted by p_u as unknown quantities. These parameters are always employed in connection with a model input u. For example, the second column in Table 4.1 defines the experimental inflow of CERT via ectopic expression $V_{32} = \tilde{p}_{u_4} u_4$. If at the beginning of an experiment ($t = 0h$) an ectopic FLAG-CERT expression is initiated, this translates *in silico* into a switch at $t = 0h$ of the input u_4 from '0' to '1' in the model. However, the effective influence of this system input on the ODE system was adjusted by estimating the scaling parameter p_{u_4} in a joint estimation with all other parameters. This input parameter estimation was also applied for all other system inputs to account for input uncertainty. Estimating input parameters for wet lab experiments and learning about the uncertainty of experimental interventions is an important point in parameter estimation. Recent findings in model calibration strongly recommend to include input uncertainty since it can prevent an underestimation of the overall uncertainty (Kaschek and Timmer, 2012).

To prepare the Bayesian model analysis the system outputs were made compatible with the measurement data. An error model with multiplicative normally distributed measurement noise (see equation (2.4) on page 24) was assumed for the final normalized data, such that for example

$$z_{\text{p-CERT, 24h}}(t) = y_{\text{p-CERT, 24h}}(t) + \eta \tag{4.19}$$

holds. Furthermore system outputs were normalized in the same way as the data, compare output equations in column three of Table 4.1 with equations in Section 4.5.

4.6.4   Design of Bayesian model analysis

After the model equations and the data set were defined we implemented a `MATLAB R2011b` script for the Bayesian model analysis. The two models were implemented in the `SBtoolbox2` model format and loaded into the `MATLAB` workspace. The whole data set including standard deviations was also loaded from a summary spreadsheet file as double precision workspace variables. Models and data are provided in the electronic supplement of Weber *et al.* (2015).

Initial parameter boundaries were chosen on a $\log_{10}$-scale as stated in Table 4.2. The displayed

Table 4.2: Upper and lower parameter boundaries, θ_i^{lb} and θ_i^{ub}, for initial MLE optimization on $\log_{10}$-scale. Units are omitted for the sake of clarity.

Parameter type	$\theta_i^{lb} \leq \theta_i \leq \theta_i^{ub}$
Parameters with linear influence:	$-5 \leq p \leq 5$
Degradation rates:	$-4 \leq a \leq 2$
Synthesis rates:	$2 \leq s \leq 7$
Input scalings:	$-1 \leq p_u \leq 8$
Threshold parameters:	$2 \leq m \leq 10$

Parameter boundaries are biologically motivated depending on the parameter type. Boundaries for parameters $\tilde{p}$ with linear influence on reaction rates were allowed to cover a wide range spanning over ten orders of magnitude. The boundaries for the degradation rates $\tilde{a}$ were estimated from typical protein half-life constants. We chose times of 30 seconds up to nine month as boundaries, which covers reasonable published values (Eden *et al.*, 2011). Lower boundaries for synthesis rates $\tilde{s}$ were set to one hundred copies per hour to guarantee some basal production of the proteins. Respective upper bounds have been set to ten million copies per hour, which we regard as a reasonable limit for the capabilities of the HEK293T protein production machinery. Input parameters $\tilde{p}_u$ add artificial production rates to the system. We did not allow them to be smaller than 10^{-1} copies per hour, otherwise we consider the ectopic expression experiment as failed. Upper limits for parameters $\tilde{p}_u$ have been set to ten million molecules per hour, assuming that experimental production rates are also limited by the cellular protein production machinery. Boundaries of reaction threshold parameters $\tilde{m}$ have been set two orders of magnitude wider than the boundaries for system variables at feasible steady states. Choosing the range for threshold parameters this wide allows to parametrize reaction kinetics f_2 (equation (4.18) on page 67) in a way to work at a saturated or non-saturated operating point. We defined system variable boundaries in equation (4.21).

To further restrict the parameter space to reasonable values for our Bayesian estimation we defined a special prior function. The initial support region of the prior is defined by the previously introduced upper and lower boundaries for the parameters, θ_i^{lb} and θ_i^{ub}. Within the support region we additionally defined restrictions for the prior distribution in form of a penalty function. It restricts the feasible steady state values of variables in the system. The prior is defined according to

$$p(\tilde{\theta}) = \begin{cases} c \cdot H(\theta) & \text{for } \theta_i \in [\theta_i^{lb}, \theta_i^{ub}] \\ 0 & \text{else} \end{cases} \tag{4.20}$$

$$\text{with } H(\theta) = \prod_{i=1}^{n_x} \left(\frac{(\bar{x}_i(\theta))^6}{(\bar{x}_i(\theta))^6 + (\bar{x}^{lb})^6} - \frac{(\bar{x}_i(\theta))^6}{(\bar{x}_i(\theta))^6 + (\bar{x}^{ub})^6} \right) \tag{4.21}$$

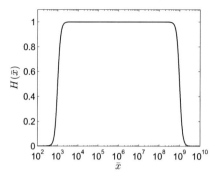

Figure 4.8: Display of the penalty function $H(\bar{x})$ used to construct the parameter prior distribution.

with steady states $\bar{x}_i(\tilde{\theta})$ and steady state lower and upper bounds $\bar{x}^{lb}$ and $\bar{x}^{ub}$. Here, c denotes a constant to correctly normalize the prior. Figure 4.8 displays the penality function $H(\bar{x})$ for the one dimensional case for a range of system steady states $\bar{x}$.

The lower and upper bounds are set to $\bar{x}^{lb} = 10^3$ and $\bar{x}^{ub} = 10^9$ molecules per cell for all species. We believe these boundaries to be very wide and not to impose a strong restriction. Furthermore, our absolute quantification experiments yielded protein abundances within these boundaries (see Figure 4.3). Literature also supports these ranges as the total amount of HEK293T proteins is estimated to $1.5 \cdot 10^{11}$ copies per cell (Finka and Goloubinoff, 2013). The sigmoid functions strongly reduce the prior probability outside the defined steady state limits. While Figure 4.8 displays $H(\bar{x})$ which is easy to compute, the dependency on the parameters $H(\theta)$ cannot be evaluated without simulating the steady states $\bar{x}_i$. However, since we have to simulate the steady state of the system anyway to fit our steady state absolute data, there is no notable extra computational cost to evaluate these implicit terms of the prior function throughout the optimization. Notably, the constant c does not necessarily need to be evaluated for a sampling run, since it cancels out in the MCMC algorithm, see equation (2.38).

To construct the likelihood function we chose a noise model based on a normal distribution, see Section 2.3, and employed empirical variance estimates for each individual data point. We employed this likelihood function and the introduced prior to construct the posterior distribution, which we subsequently used to gain initial MAP estimates. We performed repeated local optimizations with different starting conditions. As a local optimizion tool we chose the MATLAB optimizer fmincon. We set the absolute and relative tolerances to a value of 10^{-8} and the log-posterior function was passed on as objective function. Initial values were drawn uniformly within the parameter ranges of Table 4.2. After 1000 optimizations, the data fit of the 20 best MAP estimates was visually inspected to assess the models basic data fitting qualities. Subsequently, the 20 best MAP estimates were chosen to draw the starting parameters θ^0 for the subsequent sampling process. The prior function support regions were re-adjusted before the MCMC sampling procedure. For the new boundaries we took the highest and lowest found values within the 20 best MAP estimates and added (or respectively subtracted) one order of magnitude of tolerance to these values. This was done independently for each parameter. A ten temperature PT-MCMC was initiated drawing the θ^0 from the 20 best MAP estimates. The individual temperatures were chosen from a power series of five, following the recommendations of Calderhead and Girolami (2009), according to

$$T_n = \left(\frac{n}{10}\right)^5 \text{ with } n \in [1,...,10] \ . \tag{4.22}$$

All chains were run until $5 \cdot 10^5$ samples were drawn. Convergence was tested using the Gelman-Rubin statistics, see Appendix B and Figure C.2. If the $\hat{R}$-values for all temperatures and parameters fulfill $\hat{R} \leq 1.1$ the model analysis was continued otherwise more samples were drawn. Three replicate PT-MCMC sampling runs are performed for both models, where we each time redrew the initial conditions from the 20 best MAP estimates. For model comparison, thermodynamic integration was performed to estimate the marginal likelihoods and the Bayes factors, according to Section 2.11. Calculations were repeated three times using, the replicate sampling runs to estimate variances of the Bayes factors.

A pseudo code summarizing the MATLAB script of the Bayesian analysis is given in Algorithm 1. Supporting information and further statistics concerning the Bayesian analysis can be found in Appendix B.

4.6.5   Results of Bayesian data fit

In this section we discuss the results of the data fit after the MCMC runs. Since the Bayesian model fits share quite some similarities we start off discussing the basic fit properties of model A, and subsequently highlight differences between the models A and B. Figure 4.9 depicts the Bayesian data fit for model A, while the respective fit for model B is shown in Appendix Figure C.4.

Sub-figure 4.9 A depicts the steady state fits corresponding to the absolute quantification results of PKD, PI4KIIIβ and CERT. All three total abundances are well matched by model A, depicted in form of marginal probability distributions. Data is given as mean and standard deviation.

Sub-figures 4.9 B show the fit to the time series measurements of the experiment from Figure 4.5 A: ectopic expression of PI4KIIIβ with subsequent PDBu activation. Model fits are depicted differently for the dynamic experiments: We display trajectory densities and the trajectories corresponding to the MAP estimate. Additionally we depicted the evolution of the 0.5% and 99.5% percentiles over time. Red colors indicate high proababilities while yellowish colors indicate lower trajectory densities. Again, data is given as mean and standard deviation. Studying the data fit in 4.5 B, we can see that over the first $24h$ the overall PI4KIIIβ amount increases until the measured absolute amounts at $24h$ are matched. After $24h$, the PI4KIIIβ absolute abundance reaches a new quasi steady state. Signals of phospho-PKD also slightly increase throughout the first $24h$, caused by feedback in the system from the increasing PI4KIIIβ amounts. At $24h$,

input : Models A and B, absolute and relative data from wet lab measurements, prior and
likelihood function, parameter boundaries

output: MAP estimates, Posterior samples models A/B, Bayes factors, Bayesian predictions

INITIALIZE

load `SBtoolbox2` models, data and options;

for *model A, B* **do**

> **LOCAL OPTIMIZATION**
>
> $1000 \times$ `fmincon` run;
>
> save MAPs, determine top 20 values;
>
> visual inspection of MAP fits;
>
> readjust prior support region using 20 best MAPs;
>
> **PT-MCMC**
>
> **for** $i=1{:}3$ **do**
>
> > draw θ^0 from 20 best MAPs;
> >
> > run PT-MCMC, drawing $5 \cdot 10^5$ parameter samples;
> >
> > save parameters, likelihood and prior values of the samples;
>
> **end**
>
> **CONVERGENCE TEST**
>
> **if** $\hat{R} \leq 1.1$ *for all temperatures and parameters* **then**
>
> > go to **MARGINAL LIKELIHOOD**;
>
> **else**
>
> > draw additional $1 \cdot 10^5$ samples for all chains, go to **CONVERGENCE TEST**;
>
> **end**
>
> **MARGINAL LIKELIHOOD**
>
> **for** $i=1{:}3$ **do**
>
> > calculate marginal likelihood via thermodynamic integration;
>
> **end**
>
> **PREDICTIONS**
>
> calculate Bayesian predictions for scenarios of interest;

end

BAYES FACTOR

for $i=1{:}3$ **do**

> calculate Bayes factors pairwise for all models;

end

Algorithm 1: Structure of the `MATLABR2011b` script for Bayesian model analysis.

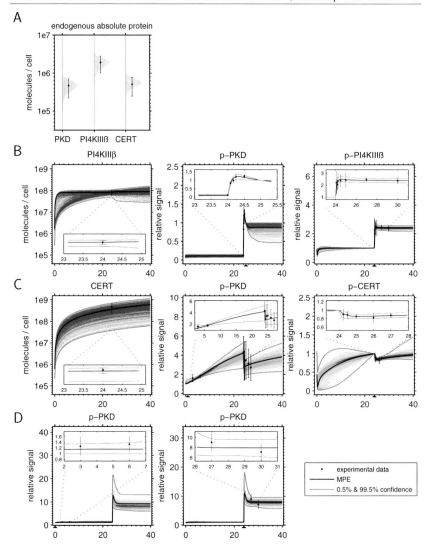

Figure 4.9: Calibration of model A to experimental data. A: Measurements of the absolute endogenous abundances of PKD, PI4KIIIβ and CERT depicted together with model steady state distributions. B: Model fit to ectopic expression of PI4KIIIβ with subsequent activation of PKD via PDBu at 24 h. Depicted are the increase of the overall PI4KIIIβ level and the response of PKD and PI4KIIIβ phosphorylation to PDBu addition. C: Model fit to ectopic expression of CERT with subsequent inhibition of PKD via kb NB 142-70 at 24 h. Depicted are the increase of the overall CERT level and the response of PKD and CERT posphorylation. D: Refinement measurements monitoring early PKD phosphorylation during ectopic expression of PI4KIIIβ and long-term PKD phosphorylation 3 and 6 hours after PDBu addition.

after addition of PDBu, phospho-PKD and phospho-PI4KIIIβ signals increase within less than ten minutes. After a slight overshoot, these proteins reach a new, increased equilibrium of their phosphorylation states at about one hour after the PDBu stimmulus.

Sub-figures 4.9 C display the model fit to the time series measurements of the experiment from Figure 4.5 B: ectopic expression of CERT with subsequent kb NB 142-70 inhibition. Total CERT levels increase to match the absolute measurements for the ectopic expressions at $24h$. Here, absolute CERT levels are still drifting to a slightly higher level after the first $24h$. Like in the previous experiment, phospho-PKD levels also increase throughout the first $24h$, since they receive feedback from the reaction network. When the inhibitor is added at $24h$, phospho-PKD and phospho-CERT levels decrease within the next hour. After $25h$ in total, phospho-PKD and phospho-CERT signals slowly recover, due to the ongoing increase in total CERT levels.

Sub-figures 4.9 D depict refinement measurements. These measurements have been planned using Bayesian experiment design methods according to Section 4.3.7 and Weber *et al.* (2012). We repeated the experiment from Sub-figure 4.9 B, with optimized measurement times. This included early measurements of phospho-PKD signals, $3h$ and $6h$ after the initiation of ectopic PI4KIIIβ expression. Additionally, we measured a long-term response of phospho-PKD $3h$ and $6h$ after addition of PDBu in an experiment that lasted over $30h$ in total. Model A very well fitted these short-term and long-term responses of phospho-PKD signals.

In summary all performed experiments can well be reproduced by model A in a holistic data fit, without any qualitative mismatch.

We now focus on the comparison with the Bayesian data fit of model B to the same data, depicted in Appendix Figure C.4. Model B also very well reproduces the steady states in Figure C.4 A and the training experiments in Figures C.4 B and C. However, there are small differences in the refinement experiments in the respective sub-figures 4.9D and C.4D. To gain a better insight we opposed both sub-figures of the short-term responses of phospho-PKD signals to PI4KIIIβ stimulation in Figure 4.10. More precisely, Figure 4.10 A displays the respective plot for model A while Figure 4.10 B shows the fit of model B. The zoom plots depict a clear discrepancy between the data and the trajectories for model B. The MAP estimate of model A matches the value of about 1.2 of the increased phospho-PKD signal. In contrast, the MAP trajectory of model B is completely flat at the basal level of 1, with no signal increase at all. The confidence intervals also indicate that model A comprises more trajectories that mimic the data then model B. The upper 99.5% confidence interval of model B does not even reach the mean of the data. There is a further difference in the early phospho-CERT levels between model A (Figure 4.9 C) and model B (Appendix Figure C.4 C). Here, model B shows a strong overshoot in the CERT phosphorylation which is not observed for model A.

4.6.6   Bayesian model comparison of models A and B

Comparing models by visually inspecting their data reproduction is important for an initial assessment of the model quality, but a real comparison requires a quantitative analysis. To

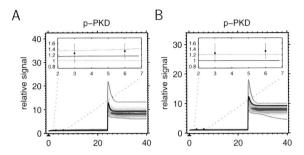

Figure 4.10: In detail comparison of Bayesian model fit. A: Bayesian data fit of model A to early response of ectopic expression of PI4KIIIβ. B: Respective data fit of model B.

gain a statistical measure for the difference in the quality of the model fits we calculated the marginal likelihoods for both models by thermodynamic integration according to Section 2.11. The results are shown in Figure 4.11. The Bayes factor is calculated by the difference of the log-marginal distributions according to equation (2.50). With a value of $2 \log(K_{A,B}) = 12.3 \pm 0.4$, the stength of evidence that model A is superior to model B is classified as 'very strong' (Kass and Raftery, 1995). In the next step we test the predictive power of the superior model A for predicting new experiments.

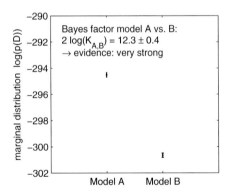

Figure 4.11: Model comparison via Bayes factors. Depicted are the log-marginal likelihoods for both models and the resulting log-Bayes factor. Standard deviations are estimated from three independent sampling runs for both models.

4.6.7   Validation of model A

To further validate model A, we used Bayesian predictions according to Section 2.9 to predict the outcome of an experiment not included in the model training. We chose one of the initial snapshot experiments from Figure 4.5 A: ectopic expression of PI4KIIIβ and inhibition of

PKD via kb NB 142-70 at $24h$. We repeated the selected snapshot experiment three times and measured the phosphorylation status of PI4KIIIβ after $25.25h$. Signals were normalized to the values measured at $24h$. The average PI4KIIIβ phosphorylation signal decreased to about 80% of the values measured after $24h$. We can see that the model very well predicts the outcome of the experiment.

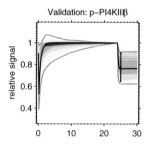

Figure 4.12: Validation experiment with data, not used in model calibration. Ectopic expression of PI4KIIIβ and inhibition of PKD via kb NB 142-70 at 24 h is performed. Measurements of phosphorylation status of PI4KIIIβ after 25.25 h are depicted together with model predictions.

In summary, the results in Sections 4.6.5, 4.6.6 and 4.6.7 support the circular reaction scheme for the ceramide transport mechanism where TGN detachment via PKD has a positive influence on the overall ceramide transfer rate. To gain a mechanistic understanding we look a bit deeper into the structure of model A. Therefore we generate Bayesian predictions based on the posterior distribution samples, for several biologically relevant scenarios.

4.6.8   Predictions I: The endogenous TGN

A biological scenario of major interest is the prediction of the endogenous state of the TGN, because it represents the natural operating mode of the average cell. As mentioned in Section 4.3.5, we mainly trained our model with data from ectopically expressed states, since many measurements were not possible in the endogenous state. The model however allows to predict the endogenous system and examine the unperturbed TGN. We repeatedly simulated the steady states of the system using parameter samples from the posterior distributions and set all experimental inputs to an inactive state. The result is depicted in form of a graph with additional information about statistics on the nodes and edges in Figure 4.13.

Colors encode the expected values of the abundances of the respective species and fluxes in molecules per cell (per h). Mini-diagrams next to the nodes additionally depict the distribution of steady state abundances including 0.5% and 99.5% percentiles.

In agreement with the measurements in Figure 4.3, PI4KIIIβ has the highest predicted abundances with about one to two million copies per cell, while PKD and CERT have about half a million proteins. Additionally, PKD and PI4KIIIβ tend to be slightly more abundant in their

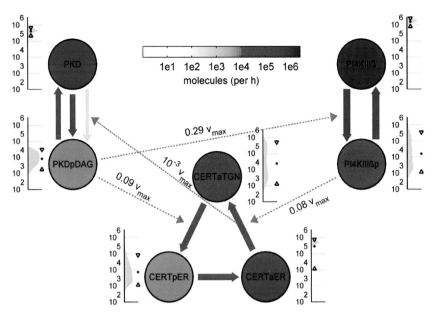

Figure 4.13: Bayesian prediction of protein amounts, fluxes and regulatory influences of the endogenous TGN. The simulation represents the endogenous TGN of an average cell. Units are given in molecules per cell. Mini diagrams next to the nodes depict the steady state distributions including expected values and 0.5% and 99.5% percentiles on a logarithmic scale. Color encoding of nodes and edges represents expected values of abundances and flow rates. Catalytic influences are represented with dashed arrows and are modeled with Michaelis Menten kinetics. Expected values of the operating points for these equations are given in units of the respective maximum reaction rate. Values below $0.5\,v_{max}$ denote responsive, non saturated couplings.

inactive form. Their active forms are present to a lesser extend and our data seems to contain less information about these states as well, since there is more uncertainty in these predictions. Reaction fluxes positively correlate with the respective protein abundances. The largest reaction rates are predicted rates for PI4KIIIβ phosphorylation and dephosphorylation followed by conversion rates related to PKD and CERT. Regarding the two phosphorylation reactions of PKD, CERT-dependent activation is predicted to be two orders of magnitude lower than the basal activation rate. The most abundant CERT representation in the model is the dephosphorylated form at the ER, while the smallest abundance is predicted for the phosphorylated form. This renders the CERT dephosphorylation process at the ER very efficient, since overall fluxes in the circular CERT reaction scheme are quite similar. Considering the predicted uncertainty of several orders of magnitude for TGN related CERT (variable 'CERTaTGN'), our data does not contain much information about TGN resting times.

For the regulatory influences we depicted the expected values of the operating status of the Michaelis Menten type kinetics, see function f_2 in equation (4.18). Operating points of regulation edges are illustrated in units of the maximal flow rate v_{max} of the respective edges. Values below $0.5v_{max}$ indicate responsive regulatory influences while edges with values above $0.5v_{max}$ tend to go into saturation upon stimulation, due to the response characteristics of the function f_2 from equation (4.18). All expected operating points of the regulatory edges are far below $0.5v_{max}$. Hence, couplings are highly responsive in the case of perturbations of the regulators and there are no saturated interactions in the endogenous system.

In summary, the endogenous TGN can be described as a system of highly responsive couplings. However, responsive single edges are no guarantee to identify advantageous regulators for biologically interesting processes, since the behavior of the closed loop feedback system must be considered. To gain a better insight regarding the interplay of PKD activity and CERT induced ceramide transfer, we further investigate system responses using selected predictions.

4.6.9   Predictions II: The role of PKD in regulating ceramide transfer

To examine the interplay between ceramide transport and PKD activation we made further predictions involving simulations of the model with slightly perturbed edges. Figure 4.14 depicts the results of these predictions.

In the first *in silico* experiment we removed the influence of the ceramide transfer between the ER and the TGN upon PKD activation. We set the respective reaction rate in model A to zero ($V_{31} = 0$), while all other system inputs remained unperturbed. Activation of PKD is now solely covered by the basal activation rate V_{12}. We plotted the relative decrease of PKD activity in Figure 4.14 A. Interestingly, the basal activation rate is able to almost completely compensate the lack of the ceramide dependent activation. Relative PKD activity decreases by less then 1%. In a second simulation, depicted in Figure 4.14 B, we increased the relative ceramide transport by 10% and monitored the relative increase of PKD activity. Again, PKD activity was not much altered and was increased by only 0.1%. These two predictions indicate that ceramide transfer activity does not much influence PKD activation and PKD activity seems to be maintained by additional processes. This behavior seems to be consistent with observations in our training data: Increasing the CERT amount to about 1000 times the basal level in the ectopic expression increases PKD activity only by a factor of four, see Figure 4.9 B. We conclude that manipulating ceramide transfer has only a minor influence on PKD activity.

If we in contrast increased PKD activity by 10% we observed an expected increase in CERT induced ceramide transport of about 6%, see 4.14 B. Hence, PKD is a candidate regulator of CERT related ceramide transport.

Summarizing the predictions, we identified PKD to be an important regulator of ceramide transfer, which is in turn not much affected by CERT.

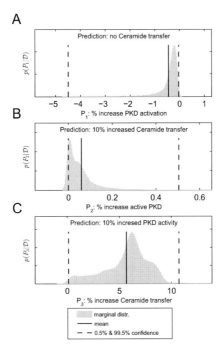

Figure 4.14: Model predictions for endogenous TGN response to perturbations in PKD activity and CERT-mediated ceramide transport: P_1: The influence of ceramide shuttling on PKD activation is set to zero in the model. Expected PKD activation drops by 2 %. P_2: The ceramide shuttle rate is increased by 10 % in the model. Expected PKD activity increases by 0.1%. P_3: PKD activity is increased by 10% in the model. Expected ceramide shuttling increases by 6%.

4.7   Summary Bayesian analysis of the TGN models

In this Chapter we presented a quantitative dynamic model describing the regulatory interactions of the TGN proteins PKD, PI4KIIIβ and CERT in HEK293T cell cultures. The phosphorylation status of these key-players influences the localization of the lipid transfer protein CERT that performs ceramide transfer between ER and Golgi membranes. The final model comprises multiple interrelated feedback loops and allows for a description of the overall network in terms of absolute abundances in molecules per cell.

For model calibration we used Bayesian model analysis and employed advanced MCMC sampling strategies. This framework allows us to analyze uncertainties in terms of probability distributions. Our calibrated ODE model was able to reproduce the signals of time series measurements of the phosphorylation states of PKD, PI4KIIIβ and CERT in a series of perturbation experiments. In the wet lab procedures we focused on perturbing the key-players by applying combinations of ectopic expression, activation and inhibitor stimuli, see Figures 4.5 and

4.9. For measurement acquisition Western Blotting techniques have been employed. To extend our data set we furthermore established an experimental procedure to estimate absolute protein abundances, see Section 4.3. We validated the predictive capabilities of the model by simulating new scenarios, which reliably agreed with the outcome of subsequent verification experiments in the wet lab.

We compared model variants that differ in the regulation of CERT mediated ceramide transfer by using Bayesian model comparison methods. The superior model (Figure 4.7 A) supported a circular reaction scheme of CERT, where the PKD induced phosphorylation of CERT establishes a positive feedback loop between PKD activity and ceramide transfer rates. In the inferior model we suggested an overall negative feedback (Figure 4.7 B) between active PKD and CERT induced ceramide transfer. When we analyzed the quality of the holistic data fit, the Bayes factor stated a very strong evidence for the cooperative behavior of CERT and PKD in form of a positive feedback, see Figure 4.11. Furthermore, we studied short-term responses of PKD activity to PI4KIIIβ expressions, where model A displayed a better reproduction of the measured signals, see Figure 4.10.

4.8   Summary, discussion and outlook

In the following section we compare our results to literature findings and reflect the advantages and limitations of the current model. We discuss the biological implications of our study and furthermore point out possible model extensions and related benefits.

4.8.1   A circular CERT reaction scheme for ceramide transfer

A major result of the model comparison study is the support of a circular reaction scheme of CERT for the ceramide transfer processes according to the biological 'short distance shuttle' model. This circular reaction scheme involves multiple phosphorylation steps of CERT to maintain ceramide transfer. From a biological perspective such a mechanism is energy-consuming, but the positive feedback structure has the benefit of a rapid response behavior in the case of an increased demand of ceramide at the TGN. Indeed, our calibrated model describes the endogenous TGN as a regulatory network that is operated in such a highly responsive state, see Section 4.6.8. The structure of model A was derived from the classical biological 'short distance shuttle' model from literature, see for example Figure 6 in Hanada (2010). The structure of model B was derived from the 'neck-swinging' mechanism. Although our study supports model A we consider it too early to entirely drop the idea of the molecular mechanism of 'neck-swinging'. However, the negative feedback regulation in the current 'neck-swinging' model must be rethought for future studies, since our study strongly supports a cooperative behavior.

4.8.2   Regulation of ER-to-TGN ceramide transfer

Increasing ceramide trafficking between the ER and the TGN is of high interest to control related processes such as the secretion rate, see Florin *et al.* (2009). Active PKD was identified to be an efficient regulator of ceramide trafficking using model simulations, see Figure 4.14. However, when we increased or decreased the ceramide transfer rate, PKD activity was not much affected. Our model predictions displayed a strong basal activation rate of PKD, which is independent of ceramide trafficking, see Figure 4.13. We conclude that the DAG pool connected to the SMS reaction has a less important role in PKD activation than initially thought. Hence, several other TGN located DAG pathways could be a target in future projects to identify the regulation mechanism responsible for the critical DAG sub-pool which regulates PKD activity and subsequently ceramide trafficking.

We introduced some of these pathways in Section 1.3.1, which allow for the synthesis or consumption of DAG without involving the SMS1 reaction, see also Sections 1.3.1.2 and 1.3.1.3. A candidate is the CDP-choline pathway which consumes DAG to form PC (Carrasco and Mérida, 2007). Alternatively, a pathway involving phospholipase D (PLD), PA phosphohydrolases (PAPs) and lipid phosphate phosphatases (LPPs) generates DAG from PC (Bard and Malhotra, 2006; Lev, 2010). Further candidate pathways involve hydrolysis of PI(4,5)P$_2$ by PLCβ3 (Díaz Añel, 2007) or dephosphorylation of phosphatidic acid by lipid-phosphate phosphatases (Shemesh *et al.*, 2003; Sarri *et al.*, 2011) to generate DAG. We argue that studying the relations between the before mentioned reaction networks and our network would be a logical step in the identification of further potential ceramide transfer regulators.

4.8.3   Measuring and predicting absolute quantities

Our model was calibrated using data including absolute quantities. This has the advantage of enabling predictions of absolute abundances. The Western Blot based method to estimate these absolute abundances was invented and validated throughout the project (see Section 4.3.4) and is still at an early state of development. Although we were able to achieve good initial results comprising estimates that deviated less then a factor of two from commercial protein quantification assays, these results must still be handled with care. We identified PI4KIIIβ to be the most abundant protein with about 1-2 million molecules in the average HEK293T cell. The abundance of PKD and CERT was estimated to about half a million proteins per cell. To support our findings we did a literature search about absolute abundances of other ubiquitous HEK293T pathways. Although there are not much references, an investigation of the HEK293T Wnt-pathway yielded estimates of average abundances of $1.6 \cdot 10^5$ copies per cell for the involved signaling proteins (Tan *et al.*, 2012). These literature values are in the same order of magnitude as our average abundances of the key proteins for the endogenous system, which were estimated to be a little less then a million copies per cell.

If we in turn predict absolute abundances with our model the accuracy strongly depends on the type of output and the scenario. While the steady state amount for e.g. inactive PKD (x_1) can be predicted with a variation of less than $3 \cdot 10^5$ molecules per cell, active PI4KIIIβ (x_4) varies over three orders of magnitude for the same scenario, see Figure 4.13. However, the Bayesian framework allows to quantify uncertainties according to Section 2.10 which allows for an individual assessment of the potential and limitations for each prediction which is a great advantage.

4.8.4   Model extensions

Connecting and extending existing models is a major future goal on the road towards a holistic picture of the secretory pathway in mammalian cells.

If we consider model extensions, an initial goal could be to integrate TGN lipids. Including the lipids Ceramide, PC, DAG and SM from the SMS1 reaction studied in Chapter 3 could help to investigate TGN related protein-lipid interactions and would establish a connection to the sphingolipid pathway (Gupta *et al.*, 2011). The investigation of this model extension is motivated by the fact that the two model studies from Chapter 3 and 4 support each other with regards to the identified regulatory structures, although the data sets used are from different cell lines. The lipid-based model from Chapters 3 supports a positive feedback from DAG to Ceramide while the protein-centered model of Chapter 4 identifies active PKD as a potent up-regulator of ceramide transfer. Since DAG attracts PKD to the TGN and activates it (Maeda *et al.*, 2001; Baron and Malhotra, 2002) the feedback loop is closed and the overall picture is consistent. However, such a model extension should be supported by additional lipid data from HEK293T cells, ideally with measurements of TGN related lipid sub-pools. A further motivation to extend the model with lipid dynamics is to establish a better connection to research related to DAG induced membrane curvature (Corda *et al.*, 2002; Shemesh *et al.*, 2003) or secretion rates (Florin *et al.*, 2009). A goal could be to connect TGN protein and lipid dynamics with these effects.

A further extension could aim for adding additional lipid transfer proteins that are active at the ER-TGN MCS, such as oxysterol-binding protein OSBP. This extension would establish additional feedback mechanisms in the model, since OSBP transport activity is regulated by PKD, it transfers PI4P from the TGN to the ER (Mesmin *et al.*, 2013) and stabilizes CERT localization at the ER-TGN MCS (Nhek *et al.*, 2010). Benefits of adding additional regulation mechanisms for ceramide transfer could be improved model predictions. It also helps to reinvestigate the currently inferior 'neck-swinging' model and design new versions of it, see Section 4.8.1. Since OSBP transfers sterol from the ER to the TGN, this extension would also connect the sphingolipid pathways with sterol metabolism. In a future project, initial data could be tested by using CERT similar transfer mechanisms from this thesis for OSBP in the model, since this transfer protein also has a FFAT motif and a PH domain (Perry and Ridgway, 2005).

Chapter 5

Final summary and conclusions

In this thesis simulation models have been developed and validated to investigate the protein-lipid interaction mechanisms at trans-Golgi network of mammalian cells. In two systems biological studies, we used ordinary differential equation models to examine the interactions between the lipids involved in the SMS1 reaction at the TGN (Weber *et al.*, 2013) and the proteins involved in the regulation of non-vesicular ER-to-TGN ceramide transfer (Weber *et al.*, 2015). This included the lipids ceramide, PC, DAG and SM and the proteins PKD, PI4KIIIβ and CERT, respectively. The final goal was to create a quantitative model of this network, study competing hypotheses regarding the mechanism of ceramide transfer and identify important regulators for this process.

These issues of major biological interest were due to the lack of direct observation methods only addressable via simulation studies, which highly motivated a model based approach. In our intensive literature research we identified the most important references of isolated experimental observations that assemble the overall interaction network between the above key-players. We used these as a starting point to create our data-driven models and incorporate experimental data. Since large uncertainties in the biological measurements cause parameter identifiability problems in the ODE model calibration process, we decided to use a Bayesian model analysis framework to deal with this issue. Here we setup and adapted multiple MCMC methods to handle the high dimensional parameter estimation problems. The final framework allows for parameter estimation (Weber *et al.*, 2011), uncertainty analysis, model predictions with uncertainties, experiment design (Weber *et al.*, 2012) and fair model comparison of models with different complexity.

In our first lipid-centered model study we used literature data from macrophages and CHO cells to show that incorporating a simple linear positive feedback from DAG to ceramide in the SMS1 reaction equations greatly improves the data fit. The inclusion of this feedback reaction was motivated by our literature research which suggested that such a feedback involves ER-to-TGN ceramide transfer processes. Although the underlying ceramide transfer process was still embedded into the reaction equations in an oversimplified way, it allowed our model to correctly predict the outcome of additional ceramide overexpression and SMS1 silencing experiments.

Furthermore this study motivated to investigate the ER-to-TGN ceramide transfer mechanism in more detail.

Hence, we designed a second protein-centered model study about ER-to-TGN ceramide transfer mechanisms including the key-players PKD, PI4KIIIβ and CERT and our own tailored experiments with HEK293T cells. We invented and evaluated our own experimental scheme to generate absolute data of the above species. Furthermore we gathered time-series data of the phosphorylation states of the proteins after dynamic perturbation experiments. Bayesian model comparison showed that our data strongly supports the 'short distance shuttle' model over the 'neck-swinging' model – two competing models that are intensively discussed in literature – for CERT induced ceramide transfer. Furthermore, we generated insights of the endogenous TGN via quantitative simulations that for the first time displayed this network under natural conditions, which is of great value for basic research. Additionally, we identified active PKD as a potent ceramide transfer rate regulator, which serves as a basis for secretory pathway optimization in producer cell lines.

Important methodological side results of this thesis are a novel absolute quantification scheme for proteins (see Section 4.3.4, and Appendix of Weber *et al.* (2015)), a Bayesian experiment design method (see Section 4.3.7 and Weber *et al.* (2012)) and a maximum likelihood parameter estimation method for biological data (for compactness only described in (Weber *et al.*, 2011)).

Our biological results furthermore have broader implications in the general field of lipid transfer protein regulation. The supported 'short distance shuttle' model includes positive feedback regulations and a circular reaction scheme for CERT dependent ceramide transfer, which serve as a good starting point to investigate the regulation of other lipid transfer proteins. Especially OSBP and phosphatidylinositol 4-phosphate adaptor protein (FAPP) also possess a PH domain for PI4P dependent TGN targeting and may be regulated in a similar way. Our results might in future impact on biotechnological applications, where targeted interventions are used to optimize the secretory pathway in producer cell lines in batch-reactors (Florin *et al.*, 2009). Currently our models are calibrated with HEK293T data, a cell line which is not a typical batch-reactor producer cell line. A further limiting factor is that our models currently feature no variables such as the 'production rate', a topic which is discussed in Section 4.8.4. On the positive side, these are feasible extensions since there are already HEK293-6E suspension cells in usage for antibody production (Jager *et al.*, 2013) and other methods for direct usage of HEK293T-derived cells for production of therapeutics (Guy *et al.*, 2013). Furthermore, our model structures were derived from results of slightly differing HEK293(T) cell types and our final models subsequently agreed with our specific HEK293T cell data. Thus, we already demonstrated the transferability potential of the model structures across various HEK293 cell types when creating our models. Results should also partly apply to further epithelial cell lines, like HeLa cells, since here structural homology in the interactions of our key-players has been observed (Baron and Malhotra, 2002; Malhotra and Campelo, 2011).

Finally, we believe that analyzing present and future experimental results with model-based studies is an important procedure towards the holistic understanding of TGN interactions, the Golgi secretory function and the overall mammalian cell. Expanding, updating and also dropping parts of models are the key processes to achieve these goals. The quantitative computational models presented in this thesis constitute a valuable contribution in this development.

Appendix A

Statistical model and definitions

This appendix chapter is based on Gelman *et al.* (2004) and the lecture notes of 'Statistical learning and stochastic control', Prof. Nicole Radde and Prof. Christian Ebenbauer, *University of Stuttgart*, 2014.

A.1 Probability space

A probability space is defined by the triple

$$(\Omega, \Sigma, P) \tag{A.1}$$

with the set of outcomes or sample space Ω, the σ-Algebra Σ and the probability measure P.

A.1.1 Set of outcomes

The non-empty set of outcomes Ω is composed of elementary outcomes

$$\Omega = \{\omega_1, \omega_2, \dots, \omega_n\}. \tag{A.2}$$

The set of outcomes can have a countable amount of elementary outcomes $n = n_x$, $|\Omega| \subseteq \mathbb{N}$ or an uncountable amount of outcomes.

A.1.2 σ-Algebra

A σ-Algebra Σ is defined as a non-empty subset $\Sigma \subset \mathcal{P}(\Omega)$ of the power set $\mathcal{P}(\Omega)$ (set of all subsets of Ω) with the following properties:

- $\emptyset, \Omega \in \Sigma$

- $A \in \Sigma \Rightarrow \bar{A} \in \Sigma$

- $A_i \in \Sigma \, \forall i \geq 1 \Rightarrow \bigcup_{i \in \mathbb{N}} A_i \in \Sigma.$

If $\mathcal{A} = \{A_1, \ldots, A_n\}$ is a non-empty set of n subsets A_i of the set of outcomes Ω, the σ-algebra generated by $\mathcal{A}$ is denoted by $\sigma(\mathcal{A})$. In the case of an uncountable set of elementary outcomes (e.g. $\Omega = \mathbb{R}$) the amount of possible subsets contained in a σ-algebra constructed on Ω is uncountable as well.

The smallest σ-algebra of Ω which contains all open sets is called Borel σ-algebra $\mathcal{B}$. The Borel σ-algebra $\mathcal{B}_{\mathbb{R}}$ of $\mathbb{R}$ is constructed by a set of intervals (a, b) such that

$$\mathcal{A}_B = \{(a, b) : a, b \in \mathbb{R}, a < b\} \tag{A.3}$$

$$\mathcal{B}_{\mathbb{R}} = \sigma(\mathcal{A}_B). \tag{A.4}$$

The elements of $\mathcal{B}_{\mathbb{R}}$ are referred to as Borel sets $B \in \mathcal{B}_{\mathbb{R}}$.

A.1.3 Measureable functions and probability measures

A measurable space is defined by the tuple

$$(\Omega, \Sigma) \tag{A.5}$$

with the set of outcomes or sample space Ω and the σ-Algebra Σ. A measure on (Ω, Σ) is defined as a function $\mu : \Sigma \to [0, \infty)$ with properties

$$\mu\left(\bigcup_{i \in \mathbb{N}} A_i\right) = \sum_{i \in \mathbb{N}} \mu(A_i) \text{ for disjoint subsets } A_i \in \Sigma, \tag{A.6}$$

$$\mu(A_i) \geq 0 \; \forall \; A_i \in \Sigma, \tag{A.7}$$

$$\mu(\emptyset) = 0. \tag{A.8}$$

The measure for the specific measure space $(\mathbb{R}, \mathcal{B}_{\mathbb{R}})$, see equations (A.3) and (A.4), is unique and reads as $\lambda = b - a$. It is referred to as the Lebesgue measure and is the most common measure used for subsets of multidimensional Euclidean spaces. All Borel sets $B \in \mathcal{B}_{\mathbb{R}}$ are Lebesgue measurable.

The Lebesgue measure of a measurable function $f : \Omega \to \mathbb{R}$ is referred to as the Lebesgue integral or simply the integral of the function according to

$$\mu(f) = \int_{\Omega} f(x)\, dx = \lambda(f). \tag{A.9}$$

The Lebesgue measure allows simplified calculus with continuous real-valued random variables and probability density functions (PDFs), introduced in the next section.

A measurable function $P : \Sigma \to [0, 1]$ is a probability measure if it fulfills

- $P(\Omega) = 1$

- $P(A) \geq 0, \; \forall A \in \Sigma$

- $A \cap B = \emptyset \Rightarrow P(A \cup B) = P(A) + P(B)$

- $P(\bigcup_{i=1}^{n} A_i) = \Sigma_{i=1}^{n} P(A_i)$ if $A_i \cap A_j = \emptyset$, $i \neq j$.

A.2   Continuous real-valued random variables (RVRVs)

Real valued random variables X are functions that map the set of outcomes to real valued numbers. Let (Ω, Σ) and $(\mathbb{R}, \mathcal{B})$ be two measurable spaces. A function $X : \Omega \to \mathbb{R}$ is called measurable real-valued random variable if X is measurable and

$$X^{-1}(B) := \{\omega \in \Omega : X(\omega) \in B\} = A \in \Sigma, \ \forall B \in \mathcal{B} \tag{A.10}$$

holds. Furthermore, if $B(\alpha) := (-\infty, \alpha]$ we define

$$X^{-1}(\alpha) := \{\omega \in \Omega : X(\omega) \leq \alpha\} = A \in \Sigma, \ \forall B(\alpha) \in \mathcal{B}. \tag{A.11}$$

Hence, random variables X assign to a Borel set B an event $A \in \Sigma$ in the σ-Algebra.

In this thesis we also employ vectors of random variables (random vectors) according to

$$X : \Omega \to \mathbb{R}^n \tag{A.12}$$

$$X^{-1}(B_n) := \{\omega \in \Omega : X(\omega) \in B_n\} = A \in \Sigma \tag{A.13}$$

with the associated measurable spaces (Ω, Σ) and $(\mathbb{R}^n, \mathcal{B}^n)$.

Furthermore,

$$X(\omega) = x \tag{A.14}$$

$$X^{-1}(x) = \omega \tag{A.15}$$

holds, for all possible elementary outcomes ω and random variates (or realizations of the random variable) x. If the domain $X(\Omega)$ is uncountable, X is referred to as a continuous random variable. A summary of the statistical foundation used in this thesis is given in Table A.1.

A RV induces a probability measure $P_X(x)$ on $\mathcal{B}$ (or $\mathcal{B}^n$) given the probability measure of the probability space (Ω, Σ, P) according to

$$P_X(x \in B) := P_X(B) = P(X^{-1}(B)) = P(\omega \in \Omega : x(\omega) \in B). \tag{A.16}$$

Furthermore, for a continuous RV there exists a probability density function $f_X(x)$ with properties

$$f_X(x) \geq 0 \ \forall x \tag{A.17}$$

$$\int_{+\infty}^{-\infty} f_X(x) dx = 1 \tag{A.18}$$

$$P(a \leq X \leq b) = \int_a^b f_X(x) \, dx, \text{ for } a \leq b \tag{A.19}$$

$$P(a < X < b) = P(a \leq X \leq b). \tag{A.20}$$

Table A.1: Summary of the statistical foundation of this thesis.

Property	Model
event space	(Ω, Σ, P)
image space	$(\mathbb{R}^n, \mathcal{B}^n, \lambda)$
random variables	$X : \Omega \to \mathbb{R}^n$

If X is a random vector of continuous RVRVs with n entries $X_1 \ldots X_n$, the joint probability density function reads as

$$f_{X_1,\ldots,X_n}(x_1,\ldots,x_n) \tag{A.21}$$

$$\text{with} \int_{\mathbb{R}^n} f_{X_1,\ldots,X_n}(x_1,\ldots,x_n)\, dx_1\, \ldots\, dx_n = 1. \tag{A.22}$$

Properties A.17, A.19, A.20 apply in an analogue way.

The function

$$F_X(\alpha) := P_X(x \leq \alpha) = P_X([-\infty,\alpha)) = P(\omega \in \Omega : x(\omega) \leq \alpha) \tag{A.23}$$

is referred to as the cumulative distribution function of a RV, with properties

$$F_X(x) \in [0,1] \tag{A.24}$$

$$F_X(-\infty) = 0,\ F_X(+\infty) = 1 \tag{A.25}$$

$$F_X(x) \text{ monotonically increasing} \tag{A.26}$$

$$F_X(x_2) - F_X(x_1) = P_X(x_1 < x \leq x_2). \tag{A.27}$$

In the case of a continuous RV the cumulative distribution function is defined as

$$F_X(x) = \int_{-\infty}^{x} f_X(z)dz. \tag{A.28}$$

A.3 Important formulas for continuous RVRVs and PDFs

The expected value of a continuous RVRV is defined as

$$\mu = E(X) = \int_{-\infty}^{+\infty} x f_X(x)dx. \tag{A.29}$$

The unbiased estimator of the expected value based on n realizations x_i of the random variable X is denoted as

$$\hat{\mu} = \frac{1}{n} \sum_{i=1}^{n} x_i. \tag{A.30}$$

The variance of a continuous RVRV is defined as

$$\sigma^2 = Var(X) = \int_{-\infty}^{+\infty} (x - \mu)^2 f_X(x)dx. \tag{A.31}$$

The unbiased estimator of the variance based on n realizations x_i of the random variable X is denoted as

$$\hat{\sigma}^2 = \frac{1}{n-1} \sum_{i=1}^{n} (x_i - \hat{\mu})^2. \tag{A.32}$$

The entries C_{ij} of the covariance matrix $C = (C_{ij})_{i,j=1...m}$ of a random vector with entries X_i, X_j are defined as

$$C_{ij} = \text{Cov}(X_i, X_j) = \text{E}\left[(X_i - \mu_i)(X_j - \mu_j)\right]. \tag{A.33}$$

The unbiased estimator of the covariance matrix entries $\hat{C}_{ij}$ based on n realizations x_i^s, x_j^s of the random vector X reads as

$$\hat{C}_{ij} = \frac{1}{n-1} \sum_{s=1}^{n} \left[(x_i^s - \hat{\mu}_i)(x_j^s - \hat{\mu}_j)\right]. \tag{A.34}$$

Here, subscript indices i, j denote the dimension and superscript index s the realization. We shortly denote the empirical estimate of the covariance matrix $\hat{C}$ of a sample $\{x^s\}_{s=1}^n$ from a random vector X with n members as $\hat{C} = Cov(\{x^s\}_{s=1}^n)$.

The marginal probability density for a single attribute (sub-dimension) X_j of a continuous real-valued random vector is defined as

$$f_{X_j}(x_j) = \int_{\mathbb{R}^{n-1}} f_{X_1,...,X_n}(x_1,...,x_n)\, dx_1\, ...\, dx_{j-1}\, dx_{j+1}\, ...\, dx_n. \tag{A.35}$$

The conditional probability density function of two random vectors X, Y with joint density $f_{X,Y}(x,y)$ reads as

$$f_{Y|X}(y|X = x) = \frac{f_{X,Y}(x,y)}{f_X(x)}, \tag{A.36}$$

where $f_Y(y|X = x)$ denotes the distribution of Y given the observation $X = x$.

Accordingly, the Bayes' theorem for probability density functions of two random vectors X, Y with joint distribution $f_{X,Y}(x,y)$ reads as

$$f_{X|Y}(x|Y = y) = \frac{f_{Y|X}(y|X = x) f_X(x)}{f_Y(y)}. \tag{A.37}$$

Appendix B

Modeling sphingomyelin synthase 1 driven conversions at the TGN

All computations have been performed using MATLAB R2011b (64 bit). For the implementation and simulation of the ODE models the SBtoolbox2 (Schmidt and Jirstrand, 2006) has been used. These toolboxes employ the ODE solver 'CVODE' from the SUNDIALS suite Cohen and Hindmarsh (1996) (http://www.llnl.gov/CASC/sundials/). Tolerances for absolute and relative error for the solver have been set to option.abstol=1e-6 and option.reltol=1e-6. To account for stiffness of the solution for some parameterizations of the ODE system, the number of maximal integration steps has been increased to options.maxnumsteps=1e6.

For parameter estimation we first used maximum likelihood estimates (MLE) to assess initial model fitting properties. To formulate the likelihood function we assumed a log-normal error model for the experimental data and made use of the empirical estimates for the standard error in the Tables 3.1 and 3.2 in Section 3.1. To generate several reliable maximum likelihood estimates we performed a multi-start optimization scheme. We employed the local optimizer fmincon with error tolerance settings for function values and constraint violations of OPTIONSfmincon.TolFun=1e-6 and OPTIONSfmincon.TolCon=1e-6. Parameter estimation and subsequent MCMC sampling was performed on logarithmic parameter space.

For the MCMC sampling we used a bounded log uniform prior. The prior support region was centered around the previously found maximum likelihood estimates. For the MCMC sampling the MATLAB coded objective function was anonymously passed on to the MCMC sampling function mcmcstat of the toolbox of Haario et al. (2006). For each of the models, two independent, parallel MCMC chains have been sampled. Firstly, a warm-up run to tune the covariance based proposal distribution was performed. For this purpose, we drew a sample with the size of 10^4 parameter vectors using the option 'DRAM' of the mcmcstat toolbox, for details see Haario et al. (2006). Subsequently we initiated the main sampling runs. For both models two chains with a total sample size of 3×10^6 parameter vectors were generated. Convergence of the chains has been tested with the Geweke method Brooks and Roberts (1998), available as an included function of the mcmcstat toolbox. All Markov chains passed the convergence analysis with a p-value of at least 0.8, testing all sub-dimensions of the parameter vectors. Subsequently, parallel chains have been merged (now featuring 6 million samples) and convergence was tested again. Resulting p-values of 0.8 or higher were calculated in each sub-dimension. The acceptance rate of the sampling runs was about about 32%.

Here we give a short summary of the numerical analysis that was performed for each of the models.

1. Initialize the SBtoolbox2 model file and load the data.

2. Run a multi-start optimizations scheme using the MATLAB fmincon optimizer to gain initial MLEs and data fits.

3. Design boundaries of the log-uniform prior based on the found MLEs.

4. Run a warm-up sampling with two parallel Markov chains to tune the proposal distribution.

5. Run a main MCMC run with parallel chains.

6. Convergence analysis and chain merging.

7. Computation of Bayesian predictions for all scenarios.

Appendix C

Modeling interactions of the TGN key-players PKD, PI4KIIIβ and CERT

C.1 PT-MCMC statistics

Figure C.1 depicts the convergence statistics of the Gelman Rubin test for the PT-MCMC runs (Gelman and Rubin, 1992) performed with model A and B from the according to Algorithm 1 on page 74.

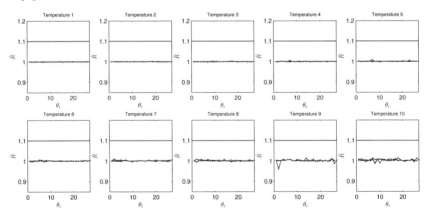

Figure C.1: MCMC convergence results. For all temperatures three population MCMC runs with different starting values were performed. Chains were tested according to the Gelman Rubin statistics. Black and blue lines denote results for Model A and B, respectively. All parameters are below the recommended value $\hat{R} \leq 1.1$ for each temperature.

Figure C.2 depicts the acceptance rate statistics for a representative PT-MCMC run of model A. Displayed are the average intra-chain acceptance rate α for the standard Metropolis steps (green), the average temperature swap acceptance rate α_t (blue) and the percentage of the initial covariance scaling factor γ (red) for all ten temperatures.

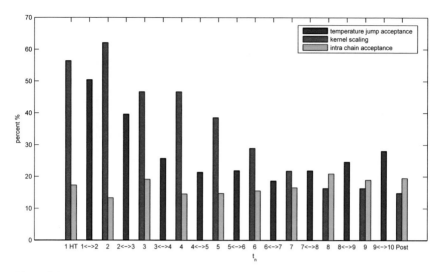

Figure C.2: Acceptance statistics of a representative run. Shown are the intra-chain aceptance rate for the Metropolis step and the percentage of the kernel scaling factor for each temperature, and temperature jump acceptance rates between temperatures.

Figure C.3 displays representative PT-MCMC samples from the analysis of model A of the temperature T_8 and the posterior distribution (T_{10}). Depicted are sub-dimensions of the full parameter vector θ. The parameters describe the degradation rates of phosphorylated and dephosphorylated PKD (here $\theta_6 = a_{11}$ and $\theta_7 = a_{12}$). The figure demonstrates that higher temperatures have a more exploratory transition kernel (see also Figure C.2). PT-MCMC facilitates sampling by flattening non-linear manifolds of the posterior distribution that host plausible parameters.

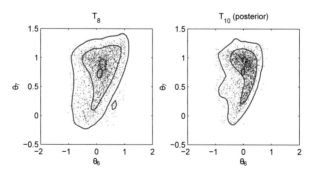

Figure C.3: Plot of samples from the PT-MCMC run of model A from the posterior T_{10} and a higher temperature T_8. Samples are represented by black dots, blue lines are level sets obtained via a KDE, red ellipsoids display the shape of the transition kernel.

C.2   Supporting results model B

Figure C.4 shows the holistic data fit for model B, compare Figure 4.9 on page 75. Model B shows differences in the data fit for early p-CERT signals (Figure C.4C third subplot) and early p-PKD levels (Figure C.4D first subplot). For the other outputs and experiments model B displays similar trajectory courses. For a systematic comparison of the model fitting qualities we refer to Section 4.6.6.

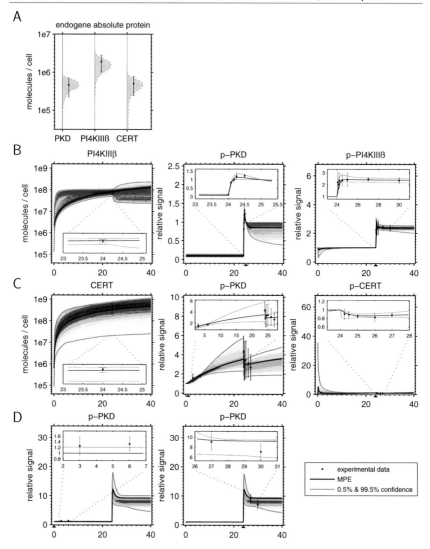

Figure C.4: Calibration of model B to experimental data. A: Measurements of the absolute endogenous abundances of PKD, PI4KIIIβ and CERT depicted together with model steady state distributions. B: Model fit to ectopic expression of PI4KIIIβ with subsequent activation of PKD via PDBu at 24 h. Depicted are the increase of the overall PI4KIIIβ level and the response of PKD and PI4KIIIβ posphorylation to PDBu addition. C: Model fit to ectopic expression of CERT with subsequent inhibition of PKD via kb NB 142-70 at 24 h. Depicted are the increase of the overall CERT level and the and the response of PKD and CERT posphorylation. D: Refinement measurements monitoring early PKD phosphorylation during ectopic expression of PI4KIIIβ and long-term PKD phosphorylation 3 and 6 hours after PDBu addition.

Bibliography

Alberts, B. (2004). *Molekularbiologie der Zelle*. Wiley-VCH.

Alpy, F. and Tomasetto, C. (2005). Give lipids a START: the StAR-related lipid transfer (START) domain in mammals. *J Cell Sci*, **118**(Pt 13), 2791–2801.

Andrieu, C. and Moulines, E. (2006). On the ergodicity properties of some adaptive MCMC algorithms. *Ann Appl Probab*, **16**(3), 1462–1505.

Añel, A. M. D. and Malhotra, V. (2005). PKCeta is required for $\beta 1\gamma 2/\beta 3\gamma 2$- and PKD-mediated transport to the cell surface and the organization of the Golgi apparatus. *J Cell Biol*, **169**(1), 83–91.

Bard, F. and Malhotra, V. (2006). The formation of TGN-to-Plasma-Membrane transport carriers. *Annu Rev Cell Dev Biol*, **22**(1), 439–455.

Baron, C. L. and Malhotra, V. (2002). Role of diacylglycerol in PKD recruitment to the TGN and protein transport to the plasma membrane. *Science*, **295**(5553), 325–328.

Blott, E. J. and Griffiths, G. M. (2002). Secretory lysosomes. *Nat Rev Mol Cell Biol*, **3**(2), 122–131.

Bravo-Altamirano, K., George, K. M., Frantz, M.-C., Lavalle, C. R., Tandon, M., Leimgruber, S., Sharlow, E. R., Lazo, J. S., Wang, Q. J., and Wipf, P. (2011). Synthesis and structure-activity relationships of benzothienothiazepinone inhibitors of protein kinase D. *ACS Med Chem Lett*, **2**(2), 154–159.

Brooks, S. P. and Roberts, G. O. (1998). Assessing convergence of Markov chain Monte Carlo algorithms. *Stat Comput*, **8**(4), 319–335.

Burger, K. N. (2000). Greasing membrane fusion and fission machineries. *Traffic*, **1**(8), 605–613.

Calderhead, B. and Girolami, M. (2009). Estimating Bayes factors via thermodynamic integration and population MCMC. *Comput Stat Data An*, **53**, 4028–4045.

Carrasco, S. and Mérida, I. (2007). Diacylglycerol, when simplicity becomes complex. *Trends Biochem Sci*, **32**(1), 27–36.

Cohen, S. and Hindmarsh, C. (1996). Cvode, a stiff/nonstiff ODE solver in C. *Comp Phys*, **10**(2), 138–143.

Consortium, U. (2010). The universal protein resource (UniProt) in 2010. *Nucleic Acids Res*, **38**, D142–D148.

Corda, D., Carcedo, C. H., Bonazzi, M., Luini, A., and Spanò, S. (2002). Molecular aspects of membrane fission in the secretory pathway. *Cell Mol Life Sci*, **59**(11), 1819–1832.

Cornish-Bowden, A. (2004). *Fundamentals of Enzyme Kinetics*. Portland Press.

D'Angelo, G., Vicinanza, M., Campli, A. D., and Matteis, M. A. D. (2008). The multiple roles of PtdIns(4)P – not just the precursor of PtdIns(4,5)P2. *J Cell Sci*, **121**(Pt 12), 1955–1963.

D'Angelo, G., Uemura, T., Chuang, C.-C., Polishchuk, E., Santoro, M., Ohvo-Rekilä, H., Sato, T., Di Tullio, G., Varriale, A., D'Auria, S., Daniele, T., Capuani, F., Johannes, L., Mattjus, P., Monti, M., Pucci, P., Williams, R. L., Burke, J. E., Platt, F. M., Harada, A., and De Matteis, M. A. (2013). Vesicular and non-vesicular transport feed distinct glycosylation pathways in the Golgi. *Nature*, **501**(7465), 116–120.

Degasperi, A., Birtwistle, M. R., Volinsky, N., Rauch, J., Kolch, W., and Kholodenko, B. N. (2014). Evaluating strategies to normalise biological replicates of Western blot data. *PLoS One*, **9**(1), e87293.

Díaz Añel, A. M. (2007). Phospholipase C β3 is a key component in the Gβγ/PKCη/PKD-mediated regulation of trans-Golgi network to plasma membrane transport. *Biochem J*, **406**(1), 157–165.

Ding, T., Li, Z., Hailemariam, T., Mukherjee, S., Maxfield, F., Wu, M.-P., and Jiang, X.-C. (2008). SMS overexpression and knockdown: impact on cellular sphingomyelin and diagylglycerol metabolism, and cell apoptosis. *J Lipid Res*, **49**, 376–85.

Droescher, A. (1998). Camillo Golgi and the discovery of the Golgi apparatus. *Histochem Cell Biol*, **109**(5-6), 425–430.

Eden, E., Geva-Zatorsky, N., Issaeva, I., Cohen, A., Dekel, E., Danon, T., Cohen, L., Mayo, A., and Alon, U. (2011). Proteome half-life dynamics in living human cells. *Science*, **331**(6018), 764–768.

Eydgahi, H., Chen, W. W., Muhlich, J. L., Vitkup, D., Tsitsiklis, J. N., and Sorger, P. K. (2013). Properties of cell death models calibrated and compared using Bayesian approaches. *Mol Syst Biol*, **9**, 644.

Farquhar, M. G. and Palade, G. E. (1981). The Golgi apparatus (complex)-(1954-1981)-from artifact to center stage. *J Cell Biol*, **91**(3 Pt 2), 77s–103s.

Fernández-Ulibarri, I., Vilella, M., Lázaro-Diéguez, F., Sarri, E., Martínez, S. E., Jiménez, N., Claro, E., Mérida, I., Burger, K. N. J., and Egea, G. (2007). Diacylglycerol is required for the formation of COPI vesicles in the Golgi-to-ER transport pathway. *Mol Biol Cell*, **18**(9), 3250–3263.

Finka, A. and Goloubinoff, P. (2013). Proteomic data from human cell cultures refine mechanisms of chaperone-mediated protein homeostasis. *Cell Stress Chaperon*, **18**(5), 591–605.

Florin, L., Pegel, A., Becker, E., Hausser, A., Olayioye, M. A., and Kaufmann, H. (2009). Heterologous expression of the lipid transfer protein CERT increases therapeutic protein productivity of mammalian cells. *J Biotechnol*, **141**(1-2), 84–90.

Fort, G., Moulines, E., and Priouret, P. (2011). Convergence of adaptive and interacting Markov chain Monte Carlo algorithms. *Ann Stat*, **39**(6), 3262–3289.

Fugmann, T., Hausser, A., Schöffler, P., Schmid, S., Pfizenmaier, K., and Olayioye, M. (2007). Regulation of secretory transport by protein kinase D-mediated phosphorylation of the ceramide transfer protein. *J Cell Biol*, **178**, 15–22.

Funato, K. and Riezman, H. (2001). Vesicular and nonvesicular transport of ceramide from ER to the Golgi apparatus in yeast. *J Cell Biol*, **155**(6), 949–959.

Gassmann, M., Grenacher, B., Rohde, B., and Vogel, J. (2009). Quantifying Western blots: pitfalls of densitometry. *Electrophoresis*, **30**(11), 1845–1855.

Gelman, A. and Rubin, D. B. (1992). Inference from iterative simulation using multiple sequences. *Stat Sci*, **7**(4), pp. 457–472.

Gelman, A., Carlin, J., Stern, H., and Rubin, D. (2004). *Bayesian data analysis*. Texts in Statistical Science. Chapman & Hall, CRC, 2nd edition.

Glick, B. S. and Nakano, A. (2009). Membrane traffic within the Golgi apparatus. *Annu Rev Cell Dev Biol*, **25**, 113–132.

Godi, A., Pertile, P., Meyers, R., Marra, P., Tullio, G. D., Iurisci, C., Luini, A., Corda, D., and Matteis, M. A. D. (1999). ARF mediates recruitment of PtdIns-4-OH kinase-beta and stimulates synthesis of PtdIns(4,5)P2 on the Golgi complex. *Nat Cell Biol*, **1**(5), 280–287.

Goñi, F. M. and Alonso, A. (1999). Structure and functional properties of diacylglycerols in membranes. *Prog Lipid Res*, **38**(1), 1–48.

Goodnight, J., Mischak, H., Kolch, W., and Mushinski, J. F. (1995). Immunocytochemical localization of eight protein kinase C isozymes overexpressed in NIH 3T3 fibroblasts. Isoform-specific association with microfilaments, Golgi, endoplasmic reticulum, and nuclear and cell membranes. *J Biol Chem*, **270**(17), 9991–10001.

Gupta, S., Maurya, M. R., Merrill, Jr, A. H., Glass, C. K., and Subramaniam, S. (2011). Integration of lipidomics and transcriptomics data towards a systems biology model of sphingolipid metabolism. *BMC Syst Biol*, **5**, 26.

Gutenkunst, R., Waterfall, J., Casey, F., Brown, K., Myers, C., and Sethna, J. (2007). Universally sloppy parameter sensitivities in systems biology models. *PLoS Comput Biol*, **3**(10), 1871–78.

Guy, H. M., McCloskey, L., Lye, G. J., Mitrophanous, K. A., and Mukhopadhyay, T. K. (2013). Characterization of lentiviral vector production using microwell suspension cultures of HEK293T-derived producer cells. *Hum Gene Ther Methods*, **24**(2), 125–139.

Haario, H., Laine, M., Mira, A., and Saksman, E. (2006). DRAM: Efficient adaptive MCMC. *Stat and Comput*, **16**, 339–354.

Hamm, T. M. (2014). Sekretionskontrolle am trans-Golgi-Netzwerk: Erstellung und Analyse eines Differentialgleichungsmodells. *Diploma Thesis - Institut für Systemtheorie und Regelungstechnik - Universität Stuttgart*, **105**.

Hanada, K. (2006). Discovery of the molecular machinery CERT for endoplasmic reticulum-to-Golgi trafficking of ceramide. *Mol Cell Biochem*, **286**(1-2), 23–31.

Hanada, K. (2010). Intracellular trafficking of ceramide by ceramide transfer protein. *Proc Jpn Acad Ser B Phys Biol Sci.*, **86**, 426–37.

Hanada, K., Kumagai, K., Yasuda, S., Miura, Y., Kawano, M., Fukasawa, M., and Nishijima, M. (2003). Molecular machinery for non-vesicular trafficking of ceramide. *Nature*, **426**(6968), 803–809.

Hanada, K., Kumagai, K., Tomishige, N., and Kawano, M. (2007). CERT and intracellular trafficking of ceramide. *Biochim Biophys Acta*, **1771**(6), 644–653.

Hanada, K., Kumagai, K., Tomishige, N., and Yamaji, T. (2009). CERT-mediated trafficking of ceramide. *Biochim Biophys Acta*, **1791**(7), 684–691.

Hannun, Y. A. and Obeid, L. M. (2002). The ceramide-centric universe of lipid-mediated cell regulation: stress encounters of the lipid kind. *J Biol Chem*, **277**(29), 25847–25850.

Hannun, Y. A. and Obeid, L. M. (2008). Principles of bioactive lipid signalling: lessons from sphingolipids. *Nat Rev Mol Cell Biol*, **9**(2), 139–150.

Hastings, W. K. (1970). Monte Carlo sampling methods using Markov chains and their applications. *Biometrika*, **57**(1), 97–109.

Hausser, A., Storz, P., Märtens, S., Link, G., Toker, A., and Pfizenmaier, K. (2005). Protein kinase D regulates vesicular transport by phosphorylating and activating phosphatidylinositol-4 kinase IIIβ at the Golgi complex. *Nat Cell Biol*, **7**(9), 880–886.

Haynes, L. P., Sherwood, M. W., Dolman, N. J., and Burgoyne, R. D. (2007). Specificity, promiscuity and localization of ARF protein interactions with NCS-1 and phosphatidylinositol-4 kinase-IIIβ. *Traffic*, **8**(8), 1080–1092.

Hirschberg, K., Miller, C. M., Ellenberg, J., Presley, J. F., Siggia, E. D., Phair, R. D., and Lippincott-Schwartz, J. (1998). Kinetic analysis of secretory protein traffic and characterization of Golgi to plasma membrane transport intermediates in living cells. *J Cell Biol*, **143**(6), 1485–1503.

Holthuis, J. C. M. and Levine, T. P. (2005). Lipid traffic: floppy drives and a superhighway. *Nat Rev Mol Cell Biol*, **6**(3), 209–220.

Hughes, H. and Stephens, D. J. (2008). Assembly, organization, and function of the COPII coat. *Histochem Cell Biol*, **129**(2), 129–151.

Huitema, K., van den Dikkenberg, J., Brouwers, J. F. H. M., and Holthuis, J. C. M. (2004). Identification of a family of animal sphingomyelin synthases. *EMBO J*, **23**(1), 33–44.

Hussain, M. M., Jin, W., and Jiang, X.-C. (2012). Mechanisms involved in cellular ceramide homeostasis. *Nutr Metab (Lond)*, **9**(1), 71.

Hyndman, R. J. and Fan, Y. (1996). Sample quantiles in statistical packages. *Am Stat*, **50**(4), pp. 361–365.

Jager, V., Bussow, K., Wagner, A., Weber, S., Hust, M., Frenzel, A., and Schirrmann, T. (2013). High level transient production of recombinant antibodies and antibody fusion proteins in HEK293 cells. *BMC Biotech*, **13**(1), 52.

Jenkins, R. W., Canals, D., and Hannun, Y. A. (2009). Roles and regulation of secretory and lysosomal acid sphingomyelinase. *Cell Signal*, **21**(6), 836–846.

Jenkins, R. W., Canals, D., Idkowiak-Baldys, J., Simbari, F., Roddy, P., Perry, D. M., Kitatani, K., Luberto, C., and Hannun, Y. A. (2010). Regulated secretion of acid sphingomyelinase: implications for selectivity of ceramide formation. *J Biol Chem*, **285**(46), 35706–35718.

Kabanikhin, S. (2008). Definitions and examples of inverse and ill-posed problems. *J Inverse Ill-pose P*, **16**, 317.

Karp, G. (2005). *Molekulare Zellbiologie*. Springer.

Kaschek, D. and Timmer, J. (2012). A variational approach to parameter estimation in ordinary differential equations. *BMC Syst Biol*, **6**(1), 99.

Kass, R. E. and Raftery, A. E. (1995). Bayes factors. *J Am Stat Assoc*, **90**(430), pp. 773–795.

Kawano, M., Kumagai, K., Nishijima, M., and Hanada, K. (2006). Efficient trafficking of ceramide from the endoplasmic reticulum to the Golgi apparatus requires a VAMP-associated protein-interacting FFAT motif of CERT. *J Biol Chem*, **281**(40), 30279–30288.

Klipp, E., Liebermeister, W., Wierling, C., Kowald, A., Lehrach, H., and Herwig, R. (2011). *Systems Biology*. Wiley.

Knippschild, U., Gocht, A., Wolff, S., Huber, N., Löhler, J., and Stöter, M. (2005). The casein kinase 1 family: participation in multiple cellular processes in eukaryotes. *Cell Signal*, **17**(6), 675–689.

Kreutz, C., M. Bartolome Rodriguez, M., Maiwald, T., Seidl, M., Blum, H., Mohr, L., and Timmer, J. (2007). An error model for protein quantification. *Bioinformatics*, **23**(20), 2747–2753.

Kudo, N., Kumagai, K., Tomishige, N., Yamaji, T., Wakatsuki, S., Nishijima, M., Hanada, K., and Kato, R. (2008). Structural basis for specific lipid recognition by CERT responsible for nonvesicular trafficking of ceramide. *Proc Natl Acad Sci USA*, **105**(2), 488–493.

Kumagai, K., Yasuda, S., Okemoto, K., Nishijima, M., Kobayashi, S., and Hanada, K. (2005). CERT mediates intermembrane transfer of various molecular species of ceramides. *J Biol Chem*, **280**(8), 6488–6495.

Kumagai, K., Kawano, M., Shinkai-Ouchi, F., Nishijima, M., and Hanada, K. (2007). Interorganelle trafficking of ceramide is regulated by phosphorylation-dependent cooperativity between the PH and START domains of CERT. *J Biol Chem*, **282**(24), 17758–17766.

Lemmon, M. A. and Ferguson, K. M. (2000). Signal-dependent membrane targeting by pleckstrin homology (ph) domains. *Biochem J*, **350 Pt 1**, 1–18.

Lev, S. (2010). Non-vesicular lipid transport by lipid-transfer proteins and beyond. *Nat Rev Mol Cell Biol*, **11**(10), 739–750.

Levine, T. P. (2007). A lipid transfer protein that transfers lipid. *J Cell Biol*, **179**(1), 11–13.

Levine, T. P. and Munro, S. (2002). Targeting of golgi-specific pleckstrin homology domains involves both ptdins 4-kinase-dependent and -independent components. *Curr Biol*, **12**(9), 695–704.

Li, C., Donizelli, M., Rodriguez, N., Dharuri, H., Endler, L., Chelliah, V., Li, L., He, E., Henry, A., Stefan, M. I., Snoep, J. L., Hucka, M., Le Novère, N., and Laibe, C. (2010). Biomodels database: An enhanced, curated and annotated resource for published quantitative kinetic models. *BMC Syst Biol*, **4**, 92.

Litvak, V., Dahan, N., Ramachandran, S., Sabanay, H., and Lev, S. (2005). Maintenance of the diacylglycerol level in the golgi apparatus by the nir2 protein is critical for golgi secretory function. *Nat Cell Biol*, **7**(3), 225–234.

Loewen, C. J. R., Roy, A., and Levine, T. P. (2003). A conserved er targeting motif in three families of lipid binding proteins and in opi1p binds vap. *EMBO J*, **22**(9), 2025–2035.

Maeda, Y., Beznoussenko, G. V., Lint, J. V., Mironov, A. A., and Malhotra, V. (2001). Recruitment of protein kinase d to the trans-golgi network via the first cysteine-rich domain. *EMBO J*, **20**(21), 5982–5990.

Malhotra, V. and Campelo, F. (2011). PKD regulates membrane fission to generate tgn to cell surface transport carriers. *Cold Spring Harb Perspect Biol*, **3**(2).

Mark, S. D. and Gail, M. H. (1994). A comparison of likelihood-based and marginal estimating equation methods for analysing repeated ordered categorical responses with missing data: application to an intervention trial of vitamin prophylaxis for oesophageal dysplasia. *Stat Med*, **13**(5-7), 479–493.

Mesmin, B., Bigay, J., Moser von Filseck, J., Lacas-Gervais, S., Drin, G., and Antonny, B. (2013). A four-step cycle driven by pi(4)p hydrolysis directs sterol/pi(4)p exchange by the er-golgi tether osbp. *Cell*, **155**(4), 830–843.

Metropolis, N., Rosenbluth, A. W., Rosenbluth, M. N., Teller, A. H., and Teller, E. (1953). Equation of state calculations by fast computing machines. *J Chem Phys*, **21**(6), 1087–1092.

Mironov, A. and Pavelka, M. (2009). *The Golgi Apparatus: State of the art 110 years after Camillo Golgi's discovery.* Springer.

Mogelsvang, S., Marsh, B. J., Ladinsky, M. S., and Howell, K. E. (2004). Predicting function from structure: 3D structure studies of the mammalian Golgi complex. *Traffic*, **5**(5), 338–345.

Nhek, S., Ngo, M., Yang, X., Ng, M. M., Field, S. J., Asara, J. M., Ridgway, N. D., and Toker, A. (2010). Regulation of OSBP Golgi localization through protein kinase D-mediated phosphorylation. *Mol Biol Cell*, **21**(13), 2327–2337.

Olayioye, M. A. and Hausser, A. (2012). Integration of non-vesicular and vesicular transport processes at the Golgi complex by the PKD-CERT network. *Biochim Biophys Acta*, **1821**(8), 1096–1103.

Parker, P. A., Vining, G. G., Wilson, S. R., Szarka, J. L., and Johnson, N. G. (2010). The prediction properties of classical and inverse regression for the simple linear calibration problem. *J Qual Technol*, **42**(4), 332–347.

Perry, D. K., Carton, J., Shah, A. K., Meredith, F., Uhlinger, D. J., and Hannun, Y. A. (2000). Serine palmitoyltransferase regulates de novo ceramide generation during etoposide-induced apoptosis. *J Biol Chem*, **275**(12), 9078–9084.

Perry, R. J. and Ridgway, N. D. (2005). Molecular mechanisms and regulation of ceramide transport. *Biochim Biophys Acta*, **1734**(3), 220–234.

Pusapati, G. V., Krndija, D., Armacki, M., von Wichert, G., von Blume, J., Malhotra, V., Adler, G., and Seufferlein, T. (2010). Role of the second cysteine-rich domain and Pro275 in protein kinase D2 interaction with ADP-ribosylation factor 1, trans-Golgi network recruitment, and protein transport. *Mol Biol Cell*, **21**(6), 1011–1022.

Raue, A., Schilling, M., Bachmann, J., Matteson, A., Schelker, M., Schelke, M., Kaschek, D., Hug, S., Kreutz, C., Harms, B. D., Theis, F. J., Klingmüller, U., and Timmer, J. (2013). Lessons learned from quantitative dynamical modeling in systems biology. *PLoS One*, **8**(9), e74335.

Rozengurt, E., Rey, O., and Waldron, R. T. (2005). Protein kinase D signaling. *J Biol Chem*, **280**(14), 13205–13208.

Rykx, A., Kimpe, L. D., Mikhalap, S., Vantus, T., Seufferlein, T., Vandenheede, J. R., and Lint, J. V. (2003). Protein kinase D: a family affair. *FEBS Lett*, **546**(1), 81–86.

Saito, S., Matsui, H., Kawano, M., Kumagai, K., Tomishige, N., Hanada, K., Echigo, S., Tamura, S., and Kobayashi, T. (2008). Protein phosphatase 2Cε is an endoplasmic reticulum integral membrane protein that dephosphorylates the ceramide transport protein CERT to enhance its association with organelle membranes. *J Biol Chem*, **283**(10), 6584–6593.

Sarri, E., Sicart, A., Lázaro-Dieǵuez, F., and Egea, G. (2011). Phospholipid synthesis participates in the regulation of diacylglycerol required for membrane trafficking at the Golgi complex. *J Biol Chem*, **286**(32), 28632–43.

Schmidt, H. and Jirstrand, M. (2006). Systems biology toolbox for matlab: a computational platform for research in systems biology. *Bioinformatics*, **22**(4), 514–515.

Shemesh, T., Luini, A., Malhotra, V., Burger, K. N. J., and Kozlov, M. M. (2003). Prefission constriction of Golgi tubular carriers driven by local lipid metabolism: a theoretical model. *Biophys J*, **85**(6), 3813–3827.

Silverman, B. (1986). *Density Estimation for Statistics and Data Analysis*. Chapman & Hall/CRC Monographs on Statistics & Applied Probability. Taylor & Francis.

Subathra, M., Qureshi, A., and Luberto, C. (2011). Sphingomyelin synthases regulate proetin trafficking and secretion. *PLoS One*, **9**(9), e23644.

Tafesse, F. G., Ternes, P., and Holthuis, J. C. M. (2006). The multigenic sphingomyelin synthase family. *J Biol Chem*, **281**(40), 29421–29425.

Tafesse, F. G., Huitema, K., Hermansson, M., van der Poel, S., van den Dikkenberg, J., Uphoff, A., Somerharju, P., and Holthuis, J. C. M. (2007). Both sphingomyelin synthases SMS1 and SMS2 are required for sphingomyelin homeostasis and growth in human HeLa cells. *J Biol Chem*, **282**(24), 17537–17547.

Tan, C. W., Gardiner, B. S., Hirokawa, Y., Layton, M. J., Smith, D. W., and Burgess, A. W. (2012). Wnt signalling pathway parameters for mammalian cells. *PLoS One*, **7**(2), e31882.

Tomishige, N., Kumagai, K., Kusuda, J., Nishijima, M., and Hanada, K. (2009). Casein kinase Iγ2 down-regulates trafficking of ceramide in the synthesis of sphingomyelin. *Mol Biol Cell*, **20**(1), 348–357.

Traub, L. M. and Kornfeld, S. (1997). The trans-Golgi network: a late secretory sorting station. *Curr Opin Cell Biol*, **9**(4), 527–533.

Tóth, B., Balla, A., Ma, H., Knight, Z. A., Shokat, K. M., and Balla, T. (2006). Phosphatidylinositol 4-kinase IIIβ regulates the transport of ceramide between the endoplasmic reticulum and Golgi. *J Biol Chem*, **281**(47), 36369–36377.

van Meer, G., Voelker, D. R., and Feigenson, G. W. (2008). Membrane lipids: where they are and how they behave. *Nat Rev Mol Cell Biol*, **9**(2), 112–124.

Varki, A. and Chrispeels, M. (1999). *Essentials of Glycobiology*. Cold Spring Harbor Laboratory Press.

Villani, M., Subathra, M., Im, Y.-B., Choi, Y., Signorelli, P., del Poeta, M., and Luberto, C. (2008). Sphingomyelin synthases regulate production of diacylglycerol at the Golgi. *Biochem J*, **414**, 31–41.

Wang, Q. J. (2006). PKD at the crossroads of DAG and PKC signaling. *Trends Pharmacol Sci*, **27**(6), 317–323.

Weber, P., Hasenauer, J., Allgöwer, F., and Radde, N. (2011). Parameter estimation and identifiability of biological networks using relative data. *Proceedings of the 18th IFAC World Congress (Milano 2011)*, pages 11648–11653.

Weber, P., Kramer, A., Dingler, C., and Radde, N. (2012). Trajectory-oriented Bayesian experiment design versus Fisher A-optimal design: an in depth comparison study. *Bioinformatics*, **28**(18), i535–i541.

Weber, P., Thomaseth, C., Hamm, T., Kashima, K., and Radde, N. (2013). Modeling sphingomyelin synthase 1 driven reaction at the golgi apparatus can explain data by inclusion of a positive feedback mechanism. *J Theor Biol*, **337**, 174–180.

Weber, P., Kuritz, K., Kramer, A., Allgöwer, F., Olayioye, M. A., Hausser, A., and Radde, N. (2014). Mit Simulationstechnik zu neuen Erkenntnissen in der Systembiologie. *10. Ausgabe des Themenheft Forschung der Universität Stuttgart*, **10**, 32–44.

Weber, P., Hornjik, M., Olayioye, M. A., Hausser, A., and Radde, N. (2015). A computational model of PKD and CERT interactions at the trans-Golgi network of mammalian cells. *Accepted - BMC Systems Biology*.

Wilkinson, D. (2006). *Stochastic modelling for systems biology*, volume 11 of *Mathematical and Computational Biology*. Chapman & Hall/CRC.